METHODS IN MOLECULAR BIOLOGY

Series Editor
John M. Walker
School of Life and Medical Sciences
University of Hertfordshire
Hatfield, Hertfordshire, AL10 9AB, UK

For further volumes:
http://www.springer.com/series/7651

Gap Junction Protocols

Editors by

Mathieu Vinken

Department of in Vitro Toxicology and Dermato-Cosmetology, Vrije Universiteit Brussel, Brussels, Belgium

Scott R. Johnstone

BHF Glasgow Cardiovascular Research, University of Glasgow, Glasgow, United Kingdom

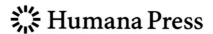 Humana Press

Editors
Mathieu Vinken
Department of in Vitro Toxicology and
 Dermato-Cosmetology
Vrije Universiteit Brussel
Brussels, Belgium

Scott R. Johnstone
BHF Glasgow Cardiovascular Research
University of Glasgow
Glasgow, United Kingdom

ISSN 1064-3745   ISSN 1940-6029   (electronic)
Methods in Molecular Biology
ISBN 978-1-4939-3662-5   ISBN 978-1-4939-3664-9   (eBook)
DOI 10.1007/978-1-4939-3664-9

Library of Congress Control Number: 2016938511

This Humana Press imprint is published by Springer Nature
The registered company is Springer Science+Business Media LLC New York

Preface

A major hallmark of multicellularity is the ability to communicate. In fact, the maintenance of homeostasis in multicellular organisms is governed by three major communicative networks, namely extracellular, intracellular, and intercellular mechanisms. Extracellular signals, like hormones, trigger intracellular messengers, mainly signal transduction mediators, which affect intercellular communication [1, 2]. The latter is mediated by gap junctions, organized in aggregates called plaques at the plasma membrane surface. Gap junctions arise from the head-to-head interaction of two hemichannels, also called connexons, of adjacent cells. These hemichannels, in turn, are built up by six connexin (Cx) proteins. More than 20 connexin species have been cloned from rodents and human, all of which are named based upon their molecular weight predicted by cDNA sequencing expressed in kilodaltons. Nevertheless, connexins share a similar structure consisting of four transmembrane domains, two extracellular loops, one cytoplasmic loop, one cytoplasmic *N*-terminal area, and one *C*-terminal region (Fig. 1). Connexins are expressed in a cell-specific way, with Cx43 being among the most widely distributed [3].

Gap junctions provide a pathway for direct communication between neighboring cells. The flux of substances through these channels is called gap junctional intercellular communication (GJIC) and concerns the passive diffusion of small (i.e. less than 1.5 kDa) and hydrophilic molecules, such as adenosine triphosphate, cyclic adenosine monophosphate, inositol triphosphate, glucose, glutathione, glutamate, as well as ions, including calcium, potassium, and sodium (Fig. 1) [4]. As numerous physiological processes are driven by messengers that are intercellularly exchanged via gap junctions, GJIC is considered to be a key mechanism in the maintenance of tissue homeostasis. GJIC indeed is involved in virtually all aspects of the cellular life cycle, ranging from cell growth to cell death [2, 5, 6]. Gap junctions equally underlie several organ-specific functions, such as in the heart, where they form an electrical syncytium [7], or in the liver, in which they control metabolic cooperation [8].

GJIC can be controlled at several levels, each typified by different kinetic patterns. Long-term GJIC regulation mainly concerns control of connexin expression. At the most upstream platform, this process is dictated by classical *cis/trans* mechanisms, whereby both ubiquitous and tissue-specific transcription factors are involved [9]. Connexin expression also relies on epigenetic mechanisms, including reversible histone modifications, DNA methylation, and microRNA-related actions [9, 10]. Short-term GJIC control is called gating and typically depends on posttranslational connexin modifications. With the exception of Cx26, all connexins are phosphoproteins [11, 12]. Gating of gap junctions is also controlled by several connexin interacting partners, such as other junctional proteins, structural proteins, and enzymes [13]. Aberrant connexin expression or GJIC regulation is associated with a plethora of pathologies, such as cancer [14], diabetes [15], cardiovascular diseases [16], and neurological disorders [17].

The present book provides a state-of-the-art compilation of protocols to study gap junctions practically. The first part describes general methods for investigating connexin expression at the transcriptional (*see* Chapter 1), epigenetic (*see* Chapter 2), and translational level (*see* Chapter 3), including techniques to examine connexin subcellular localization (*see* Chapter 4), as well as tools to experimentally modify connexin expression in vitro

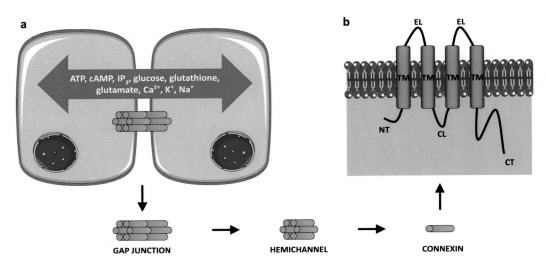

Fig. 1 (**a**) Gap junctions are built up by two hemichannels of adjacent cells, in turn composed of six connexin proteins. Gap junctions mediate direct communication between adjacent cells by controlling the intercellular diffusion of molecules like adenosine trisphosphate (ATP), cyclic adenosine monophosphate (cAMP), inositol trisphosphate (IP_3), glucose, glutathione, glutamate, as well as ions, including calcium (Ca^{2+}), potassium (K^+), and sodium (Na^+). (**b**) Connexin proteins have a common structure, consisting of four transmembrane domains (TM), two extracellular loops (EL), one cytosolic loop (CL), one cytosolic aminotail (NT), and one cytosolic carboxytail (CT)

(*see* Chapter 5) and in vivo (*see* Chapter 6). Analyses of posttranslational connexin modifications (*see* Chapter 7) and connexin interactions (*see* Chapter 8), being major determinants of gap junction gating, are also outlined.

In the second part of the book, a number of established and more recently introduced approaches to functionally probe GJIC are presented. The majority of these methods are based upon monitoring of the intercellular exchange of reporter dyes, brought into cells by mechanical scraping (*see* Chapter 9), microinjection (*see* Chapter 10), or electroporation (*see* Chapter 11), or variants of those involving local quenching (*see* Chapter 12) or activation (*see* Chapter 13) of fluorescent indicators. Additional GJIC protocols included in this book focus on measurement of the actual physiological functions of gap junctions, including mediating passage of secondary messengers between cells (*see* Chapter 14), facilitating intercellular calcium wave propagation (*see* Chapter 15), and providing electrical conductance (*see* Chapter 16).

The current book is intended for basic and applied researchers, ranging from the undergraduate to the postdoctoral and professional level, in the area of biomedical and life sciences, both in academic and industrial settings. It can be used by investigators familiar and unfamiliar with the gap junction field.

At the start of this book, the editors would like to express their deepest gratitude to all chapter contributors. Furthermore, the editors greatly acknowledge the Springer team and, in particular, series editor John M. Walker for his continuous assistance during the preparation of this book.

Brussels, Belgium   *Mathieu Vinken*
Glasgow, UK   *Scott R. Johnstone*

References

1. Trosko JE, Ruch RJ (2002) Gap junctions as targets for cancer chemoprevention and chemotherapy. Curr Drug Targets 3:465–482
2. Vinken M, Vanhaecke T, Papeleu P et al (2006) Connexins and their channels in cell growth and cell death. Cell Signal 18:592–600
3. Bai D, Wang AH (2014) Extracellular domains play different roles in gap junction formation and docking compatibility. Biochem J 458:1–10
4. Alexander DB, Goldberg GS (2003) Transfer of biologically important molecules between cells through gap junction channels. Curr Med Chem 10:2045–2058
5. Vinken M, Decrock E, De Vuyst E et al (2011) Connexins: sensors and regulators of cell cycling. Biochim Biophys Acta 1815:13–25
6. Decrock E, Vinken M, De Vuyst E et al (2009) Connexin-related signaling in cell death: to live or let die? Cell Death Differ 16:524–536
7. Veeraraghavan R, Poelzing S, Gourdie RG (2014) Intercellular electrical communication in the heart: a new, active role for the intercalated disk. Cell Commun Adhes 21:161–167
8. Vinken M, Henkens T, De Rop E et al (2008) Biology and pathobiology of gap junctional channels in hepatocytes. Hepatology 47:1077–1088
9. Oyamada M, Takebe K, Oyamada Y (2013) Regulation of connexin expression by transcription factors and epigenetic mechanisms. Biochim Biophys Acta 1828:118–133
10. Vinken M, De Rop E, Decrock E et al (2009) Epigenetic regulation of gap junctional intercellular communication: more than a way to keep cells quiet? Biochim Biophys Acta 1795:53–61
11. Johnstone SR, Billaud M, Lohman AW et al (2012) Posttranslational modifications in connexins and pannexins. J Membr Biol 245:319–332
12. D'hondt C, Iyyathurai J, Vinken M et al (2013) Regulation of connexin- and pannexin-based channels by post-translational modifications. Biol Cell 105:373–398
13. Gilleron J, Carette D, Chevallier D et al (2012) Molecular connexin partner remodeling orchestrates connexin traffic: from physiology to pathophysiology. Crit Rev Biochem Mol Biol 47:407–423
14. Brücher BL, Jamall IS (2014) Cell-cell communication in the tumor microenvironment, carcinogenesis, and anticancer treatment. Cell Physiol Biochem 34:213–243
15. Farnsworth NL, Benninger RK (2014) New insights into the role of connexins in pancreatic islet function and diabetes. FEBS Lett 588:1278–1287
16. Fontes MS, van Veen TA, de Bakker JM et al (2012) Functional consequences of abnormal Cx43 expression in the heart. Biochim Biophys Acta 1818:2020–2029
17. Dere E, Zlomuzica A (2012) The role of gap junctions in the brain in health and disease. Neurosci Biobehav Rev 36:206–217

Contents

Preface. . . . . *v*

Contributors. . . . . *xi*

1 Analysis of Liver Connexin Expression Using Reverse
   Transcription Quantitative Real-Time Polymerase Chain Reaction. 1
   Michaël Maes, Joost Willebrords, Sara Crespo Yanguas, Bruno Cogliati,
   and Mathieu Vinken

2 DNA Methylation Analysis of Human Tissue-Specific Connexin Genes 21
   Xiaoming Liu and Canxia Xu

3 Detection of Connexins in Liver Cells Using Sodium Dodecyl Sulfate
   Polyacrylamide Gel Electrophoresis and Immunoblot Analysis. 37
   Joost Willebrords, Michaël Maes, Sara Crespo Yanguas, Bruno Cogliati,
   and Mathieu Vinken

4 Immunohisto- and Cytochemistry Analysis of Connexins 55
   Bruno Cogliati, Michaël Maes, Isabel Veloso Alves Pereira, Joost Willebrords,
   Tereza Cristina Da Silva, Sara Crespo Yanguas, and Mathieu Vinken

5 Small Interfering RNA-Mediated Connexin Gene Knockdown
   in Vascular Endothelial and Smooth Muscle Cells. 71
   Miranda E. Good, Daniela Begandt, Leon J. DeLalio, Scott R. Johnstone,
   and Brant E. Isakson

6 Generation and Use of Trophoblast Stem Cells and Uterine Myocytes
   to Study the Role of Connexins for Pregnancy and Labor. 83
   Mark Kibschull, Stephen J. Lye, and Oksana Shynlova

7 Identification of Connexin43 Phosphorylation and *S*-Nitrosylation
   in Cultured Primary Vascular Cells 97
   Alexander W. Lohman, Adam C. Straub, and Scott R. Johnstone

8 Preparation of Gap Junctions in Membrane Microdomains
   for Immunoprecipitation and Mass Spectrometry Interactome Analysis 113
   Stephanie Fowler, Mark Akins, and Steffany A. L. Bennett

9 Scrape Loading/Dye Transfer Assay 133
   Pavel Babica, Iva Sovadinová, and Brad L. Upham

10 Microinjection Technique for Assessment of Gap Junction Function. 145
    Michael D. Fridman, Jun Liu, Yu Sun, and Robert M. Hamilton

11 Electroporation Loading and Dye Transfer: A Safe and Robust Method
    to Probe Gap Junctional Coupling 155
    Elke Decrock, Marijke De Bock, Diego De Baere, Delphine Hoorelbeke,
    Nan Wang, and Luc Leybaert

12 Using Fluorescence Recovery After Photobleaching to Study Gap
    Junctional Communication In Vitro. 171
    Maria Kuzma-Kuzniarska, Clarence Yapp, and Philippa A. Hulley

13 Tracking Dynamic Gap Junctional Coupling in Live Cells
    by Local Photoactivation and Fluorescence Imaging.. 181
    Song Yang and Wen-Hong Li

14 A Functional Assay to Assess Connexin 43-Mediated Cell-to-Cell
    Communication of Second Messengers in Cultured Bone Cells 193
    Joseph P. Stains and Roberto Civitelli

15 Calcium Wave Propagation Triggered by Local Mechanical Stimulation
    as a Method for Studying Gap Junctions and Hemichannels 203
    Jegan Iyyathurai, Bernard Himpens, Geert Bultynck, and Catheleyne D'hondt

16 Establishment of the Dual Whole Cell Recording Patch Clamp
    Configuration for the Measurement of Gap Junction Conductance 213
    Richard D. Veenstra

Index *233*

Contributors

MARK AKINS • *Neural Regeneration Laboratory, Department of Biochemistry, Microbiology, and Immunology, University of Ottawa, Ottawa, ON, Canada*

PAVEL BABICA • *Research Centre for the Toxic Compounds in the Environment (RECETOX), Masaryk University, Brno, Czech Republic*

DIEGO DE BAERE • *Physiology Group, Department of Basic Medical Sciences, Ghent University, Ghent, Belgium*

DANIELA BEGANDT • *Robert M. Berne Cardiovascular Research Center, University of Virginia School of Medicine, Charlottesville, VA, USA*

STEFFANY A.L. BENNETT • *Neural Regeneration Laboratory, Department of Biochemistry, Microbiology, and Immunology, University of Ottawa, Ottawa, ON, Canada*

MARIJKE DE BOCK • *Physiology Group, Department of Basic Medical Sciences, Ghent University, Ghent, Belgium*

GEERT BULTYNCK • *Laboratory of Molecular and Cellular Signaling, Department Cellular and Molecular Medicine, Katholieke Universiteit Leuven, Leuven, Belgium*

ROBERTO CIVITELLI • *Division of Bone and Mineral Diseases, Department of Medicine, Washington University, St. Louis, MO, USA*

BRUNO COGLIATI • *Department of Pathology, School of Veterinary Medicine and Animal Science, University of São Paulo, São Paulo, SP, Brazil*

CATHELEYNE D'HONDT • *Laboratory of Molecular and Cellular Signaling, Department Cellular and Molecular Medicine, Katholieke Universiteit Leuven, Leuven, Belgium*

ELKE DECROCK • *Physiology Group, Department of Basic Medical Sciences, Ghent University, Ghent, Belgium*

LEON J. DELALIO • *Robert M. Berne Cardiovascular Research Center, University of Virginia School of Medicine, Charlottesville, VA, USA; Department of Pharmacology, University of Virginia School of Medicine, Charlottesville, VA, USA*

STEPHANIE FOWLER • *Neural Regeneration Laboratory, Department of Biochemistry, Microbiology, and Immunology, University of Ottawa, Ottawa, ON, Canada*

MICHAEL D. FRIDMAN • *Department of Physiology and Laboratory Medicine, Hospital for Sick Children, Toronto, ON, Canada*

MIRANDA E. GOOD • *Robert M. Berne Cardiovascular Research Center, University of Virginia School of Medicine, Charlottesville, VA, USA*

ROBERT M. HAMILTON • *Department of Physiology and Laboratory Medicine, Hospital for Sick Children, Toronto, ON, Canada*

BERNARD HIMPENS • *Laboratory of Molecular and Cellular Signaling, Department of Cellular and Molecular Medicine, Katholieke Universiteit Leuven, Leuven, Belgium*

DELPHINE HOORELBEKE • *Physiology Group, Department of Basic Medical Sciences, Ghent University, Ghent, Belgium*

PHILIPPA A. HULLEY • *Nuffield Department of Orthopaedics, Rheumatology, and Musculoskeletal Sciences, University of Oxford, Oxford, UK*

BRANT E. ISAKSON • *Robert M. Berne Cardiovascular Research Center, University of Virginia School of Medicine, Charlottesville, VA, USA; Department of Molecular Physiology and Biophysics, University of Virginia School of Medicine, Charlottesville, VA, USA*

JEGAN IYYATHURAI • *Laboratory of Molecular and Cellular Signaling, Department of Cellular and Molecular Medicine, Katholieke Universiteit Leuven, Leuven, Belgium*

SCOTT R. JOHNSTONE • *Institute of Cardiovascular and Medical Sciences, College of Medical, Veterinary, and Life Sciences, University of Glasgow, Glasgow, UK*

MARK KIBSCHULL • *Lunenfeld-Tanenbaum Research Institute, Mount Sinai Hospital Toronto, Toronto, ON, Canada*

MARIA KUZMA-KUZNIARSKA • *Nuffield Department of Orthopaedics, Rheumatology, and Musculoskeletal Sciences, University of Oxford, Oxford, UK*

LUC LEYBAERT • *Physiology Group, Department of Basic Medical Sciences, Ghent University, Ghent, Belgium*

WEN-HONG LI • *Department of Cell Biology, University of Texas Southwestern Medical Center, Dallas, TX, USA; Department of Biochemistry, University of Texas Southwestern Medical Center, Dallas, TX, USA*

JUN LIU • *Department of Mechanical and Industrial Engineering, University of Toronto, Toronto, ON, Canada*

XIAOMING LIU • *Department of Gastroenterology, Third Xiangya Hospital of Central South University, Changsha, Hunan, China*

ALEXANDER W. LOHMAN • *Department of Cell Biology and Anatomy, Hotchkiss Brain Institute, University of Calgary, Calgary, AB, Canada*

STEPHEN J. LYE • *Lunenfeld-Tanenbaum Research Institute, Mount Sinai Hospital Toronto, Toronto, ON, Canada*

MICHAËL MAES • *Department of In Vitro Toxicology and Dermato-Cosmetology, Vrije Universiteit Brussel, Brussels, Belgium*

ISABEL VELOSO ALVES PEREIRA • *Department of Pathology, School of Veterinary Medicine and Animal Science, University of São Paulo, São Paulo, SP, Brazil*

OKSANA SHYNLOVA • *Lunenfeld-Tanenbaum Research Institute, Mount Sinai Hospital Toronto, Toronto, ON, Canada*

TEREZA CRISTINA DA SILVA • *Department of Pathology, School of Veterinary Medicine and Animal Science, University of São Paulo, São Paulo, SP, Brazil*

IVA SOVADINOVÁ • *Research Centre for the Toxic Compounds in the Environment (RECETOX), Masaryk University, Brno, Czech Republic*

JOSEPH P. STAINS • *Department of Orthopaedics, University of Maryland School of Medicine, Baltimore, MD, USA*

ADAM C. STRAUB • *Department of Pharmacology and Chemical Biology, Heart, Lung, Blood, and Vascular Medicine Institute, University of Pittsburgh, Pittsburgh, PA, USA*

Yu Sun • *Department of Mechanical and Industrial Engineering, University of Toronto, Toronto, ON, Canada*

Brad L. Upham • *Department of Pediatrics and Human Development, Michigan State University, East Lansing, MI, USA; Institute for Integrative Toxicology, Michigan State University, East Lansing, MI, USA*

Richard D. Veenstra • *Department of Pharmacology, SUNY Upstate Medical University, Syracuse, NY, USA; Department of Cell and Developmental Biology, SUNY Upstate Medical University, Syracuse, NY, USA*

Mathieu Vinken • *Department of In Vitro Toxicology and Dermato-Cosmetology, Vrije Universiteit Brussel, Brussels, Belgium*

Nan Wang • *Physiology Group, Department of Basic Medical Sciences, Ghent University, Ghent, Belgium*

Joost Willebrords • *Department of In Vitro Toxicology and Dermato-Cosmetology, Vrije Universiteit Brussel, Brussels, Belgium*

Canxia Xu • *Department of Gastroenterology, Third Xiangya Hospital of Central South University, Changsha, Hunan, China*

Song Yang • *State Key Laboratory Breeding Base of Green Pesticide and Agricultural Bioengineering, Center for R&D of Fine Chemicals, Guizhou University, Guiyang, China*

Sara Crespo Yanguas • *Department of In Vitro Toxicology and Dermato-Cosmetology, Vrije Universiteit Brussel, Brussels, Belgium*

Clarence Yapp • *Structural Genomics Consortium, University of Oxford, Oxford, UK*

Chapter 1

Analysis of Liver Connexin Expression Using Reverse Transcription Quantitative Real-Time Polymerase Chain Reaction

Michaël Maes, Joost Willebrords, Sara Crespo Yanguas, Bruno Cogliati, and Mathieu Vinken

Abstract

Although connexin production is mainly regulated at the protein level, altered connexin gene expression has been identified as the underlying mechanism of several pathologies. When studying the latter, appropriate methods to quantify connexin RNA levels are required. The present chapter describes a well-established reverse transcription quantitative real-time polymerase chain reaction procedure optimized for analysis of hepatic connexins. The method includes RNA extraction and subsequent quantification, generation of complementary DNA, quantitative real-time polymerase chain reaction, and data analysis.

Key words Connexins, RNA extraction, Reverse transcription, Minimum Information for publication of Quantitative real-time PCR Experiments

1 Introduction

Connexin (Cx) signaling can be regulated by a plethora of mechanisms at the transcriptional, posttranscriptional, translational, and posttranslational level [1, 2]. Regarding the former, connexin expression is predominantly controlled by the conventional *cis/trans* machinery [1]. A basal level of connexin gene transcription is maintained by general transcription factors, such as specificity protein 1 and activator protein 1, while tissue-specific expression depends on cell type-specific repressors and activators, such as hepatocyte nuclear factor 1α for Cx32 expression in the liver [3, 4]. In addition, epigenetic mechanisms, including histone modifications, DNA methylation, and microRNA-related control, are essential determinants of connexin gene transcription [5, 6].

Four methods are commonly used for studying individual target gene transcription, namely northern blotting [7], in situ hybridization [8], RNAse protection assays [9, 10], and reverse transcription

Mathieu Vinken and Scott R. Johnstone (eds.), *Gap Junction Protocols*, Methods in Molecular Biology, vol. 1437,
DOI 10.1007/978-1-4939-3664-9_1, © Springer Science+Business Media New York 2016

polymerase chain reaction (RT-PCR) [11]. The main limitation of the former three techniques is their relative low sensitivity, which is not the case for RT-PCR analysis [12]. The latter has a wide dynamic range, as poor and abundant expressed genes can be detected with this technique. In contrast to the conventional RT-PCR procedure, typically followed by agarose gel electrophoresis, the read-out in reverse transcription quantitative real-time polymerase chain reaction (RT-qPCR) is monitored throughout the PCR process as such and is characterized by the reaction time, during cycling, when amplification of the target is first detected. Being a quick, accurate, sensitive, specific, and cost-effective method, RT-qPCR analysis has now become the benchmark assay for quantification of RNA [13].

The liver was the first organ in which connexins have been described [14, 15]. Hepatocytes, the main hepatic cellular population, express Cx32 and to a lesser extent Cx26. In contrast, most nonparenchymal liver cells harbor Cx43 [16, 17]. In several liver diseases, such as chronic hepatitis, cirrhosis, and hepatocellular carcinoma, connexin RNA content is altered [18]. When studying the latter, appropriate methods to quantify connexin RNA levels are required. This chapter provides a two-step RT-qPCR procedure optimized for analysis of hepatic connexins, specifically Cx26, Cx32, and Cx43. Compared to the one-step RT-qPCR procedure, where the reverse transcription and the polymerase chain reaction take place in one buffer system, this is performed in two separate systems in the two-step RT-qPCR procedure. In essence, the procedure implies RNA extraction and quantification, total RNA reverse transcription into complementary DNA (cDNA) followed by a separate amplification of the cDNA by PCR and data analysis. The protocol follows the recommendations provided in the Minimum Information for publication of Quantitative real-time PCR Experiment (MIQE) guidelines [19, 20], which is a state-of-the-art guide for all the necessary requirements for experimental setup, analysis, and publication.

2   Materials

2.1   RNA Extraction

1. GenElute™ Mammalian Total RNA Miniprep Kit (Sigma, USA) (*see* **Note 1**): lysis solution, 2-mercaptoethanol, wash solution 1 (*see* **Note 2**), wash solution 2 concentrate, elution solution, GenElute™ filtration columns in tubes, GenElute™ binding columns in tubes, and collection tubes of 2.0 mL. These reagents must be stored at room temperature.

2. Lysis solution/2-mercaptoethanol mixture: add 10 µL 2-mercaptoethanol for each 1 mL of lysis solution *ex tempore* (*see* **Note 3**).

3. ≥99.5 % anhydrous ethanol.

4. Wash solution 2: dilute 2.5 mL of the provided wash solution 2 concentrate to 10 mL with ≥99.5 % anhydrous ethanol *ex tempore*.

5. 70 % ethanol solution.

6. Vortex.

7. RNase-free pipette tips: aerosol barrier recommended.

8. RNase-free microcentrifuge tubes.

9. Microcentrifuge.

10. On-Column DNase I Digestion Set (Sigma, USA): DNase digestion buffer, DNase I, binding column, and wash solution 1. The set may be stored between 2 and 8 °C for up to 6 months. For longer term, storage at –20 °C is recommended.

11. DNase I/digest buffer mixture: mix 10 µL of DNase I with 70 µL of DNase digest buffer for each preparation. Mix by inversion. Do not vortex the DNase I or the DNase I/digest buffer mixture. The mixture may be prepared up to 2 h in advance.

12. RNA*later*™, RNA stabilization solution for tissue (Sigma, USA).

13. Rotor-stator homogenizer or RNA-free pellet mixer (VWR, USA).

2.2 RNA-DNA Quantification and Purity Control

1. NanoDrop® ND-100 Spectrophotometer (Thermo Scientific, USA).

2. RNase-free pipette tips: aerosol barrier recommended.

2.3 cDNA Synthesis

1. iCycler iQ™ (Bio-Rad, USA).

2. 96-well thin wall plates.

3. Optical sealing tape.

4. iScript™ cDNA Synthesis Kit (Bio-Rad, USA): 5× iScript™ reaction mix (*see* **Note 1**), nuclease-free water, and iScript™ reverse transcriptase. The reagents must be stored at –20 °C, except for the nuclease-free water, which can be stored at room temperature. The reagents are stable for a minimum of 1 year.

5. 0.2 mL nuclease-free tubes (Bio-Rad, USA).

2.4 cDNA Purification

1. GenElute™ PCR Clean-Up Kit (Sigma, USA): column preparation solution, binding solution, wash solution concentrate, elution solution, GenElute™ plasmid mini spin column, and collection tubes of 2.0 mL. The reagents should be stored at room temperature.

2. ≥99.5 % anhydrous ethanol.

3. Wash solution: dilute 12 mL of he provided concentrate with 48 mL of ≥99.5 % anhydrous ethanol.

4. Microcentrifuge.

5. RNase-free pipette tips: aerosol barrier recommended.

6. RNase-free microcentrifuge tubes.

7. Nuclease-free water.

2.5 Real-Time qPCR

1. TaqMan® Universal PCR Master Mix (Applied Biosystems, USA): Taqman® probe, AmpliTaq Gold® DNA polymerase, AmpErase® uracil-N-glycosylase deoxynucleotides with 2′-deoxyuridine 5′-triphosphate, passive reference, and optimized buffer components. The reagents must be stored at 2–8 °C.

2. Taqman® primer (Table 1) (Applied Biosystems, USA).

3. StepOnePlus™ real-time PCR system (Applied Biosystems, USA).

4. RNase-free microcentrifuge tubes.

5. Microcentrifuge.

6. Assay-on-Demand™ Gene Expression Assay Mix (Applied Biosystems, USA).

7. Nuclease-free water.

8. Centrifuge suitable for cooling at 4 °C with adapter for 96-well plate.

9. MicroAmp® optical 96-well reaction plates (Applied Biosystems, USA).

Table 1
Primers and probes for murine connexins and candidate reference genes

Gene symbol	Assay ID	Accession number	Assay location	Amplicon size (base pairs)	Exon boundary
Gjb2	Mm00433643_s1	NM_008125.3	603	72	2–2
Gjb1	Mm01950058_s1	NM_008124.2	466	65	1–1
Gja1	Mm01179639_s1	NM_010288.3	2937	168	2–2
18S	Hs99999901_s1	X03205.1	604	187	1–1
Actb	Mm00607939_s1	NM_007393.3	1233	115	6–6
B2m	Mm00437762_m1	NM_009735.3	111	77	1–2
Gapdh	Mm99999915_g1	NM_008084.2	265	107	2–3
Hmbs	Mm01143545_m1	NM_013551.2	473	81	6–7
Ubc	Mm02525934_g1	NM_019639.4	370	176	2–2

Assay identification (ID), accession number, assay location, amplicon size, and exon boundary of target and candidate reference mouse genes (18S, 18S ribosomal RNA; Actb, β-actin; B2m, β-2-microglobulin; Gapdh, glyceraldehyde 3-phosphate dehydrogenase; Gja1, Cx43; Gjb1, Cx32; Gjb2, Cx26; Hmbs, hydroxymethylbilane synthase; Ubc, ubiquitin C)

10. MicroAmp® optical caps (Applied Biosystems, USA).

11. MicroAmp® optical tubes (Applied Biosystems, USA).

12. Pipette tips: aerosol barrier recommended.

13. Vortex.

2.6 Data Processing

1. Reference gene validation systems, such as geNorm, NormFinder, and BestKeeper. GeNorm is currently integrated in the qbase⁺ software (Biogazelle, Belgium). The BestKeeper and NormFinder software can be downloaded from http://www.gene-quantification.de/bestkeeper.html#download and http://moma.dk/normfinder-software, respectively.

2. Normalization and relative quantification of RNA software, such as qbase⁺ and the relative expression software tool (REST) (Qiagen, USA).

3 Methods

3.1 Maintenance of a Contamination-Free Workplace

Since RNases and DNases can rapidly degrade RNA and DNA, respectively, it is of utmost importance to take measures to avoid degradation. Sample acquisition constitutes the first potential source of experimental variability due to degradation. RNA yield and quality are easily perturbed by sample collection and processing methods [21].

The following steps should be taken into consideration throughout the complete RT-qPCR procedure in order to diminish the risk of contamination:

1. A strict separation between the pre-PCR and post-PCR procedures. If this is not possible, a separate enclosure for either operation should be considered. This separation also implies different sets of reagents and equipment.

2. A process flow in a unidirectional way must be constructed, implying that the PCR setup should be performed in a template-free area with reagents that under no circumstances will come in contact with possible contamination sources.

3. Benchtop hoods with high-efficiency particulate arrestance filters may be useful.

4. Nonporous surfaces should be habitually cleaned with a 10% bleach solution.

5. A water container should not be used for long-term water storage, since some bacterial species may flourish.

6. When stock solutions of the PCR reaction mix are recurrently entered with pipettes that may be contaminated with nucleic acids, aliquots of this mix should be prepared. Thus, if contamination

is suspected, the aliquots currently in use may be discarded and replaced.

7. Sample handling should be minimized and tubes opened very carefully, preferably with a tube opener that can be easily decontaminated.

8. Gloves should be changed frequently. Nothing should be touched with bare hands.

9. Aerosol-filtered pipette tips or positive displacement pipettors should be used. Tips should never be touched by anything but the pipettor.

10. Often used equipment, such as pipettors and work surfaces, should be regularly decontaminated.

3.2 RNA Extraction

Different RNA extraction kits are commercially available (e.g., RNeasy Minikit, Qiagen, USA). In this protocol, the GenElute™ Mammalian Total RNA Miniprep Kit (Sigma, USA) is used. This kit provides a simple and convenient procedure to isolate total RNA from mammalian cells and tissues, *in casu* liver tissue. For RT-qPCR purposes, an additional purification step is needed, as even minor DNA contamination can give false positive detections. Therefore, a digestion step of DNA has been introduced to the outlined procedure using the DNase digestion set (Sigma, USA). All steps in this section are carried out at room temperature.

3.2.1 Cultured Hepatic Cells

Cells grown on cell culture dishes can be lysed in situ or pelleted and stored at –80 °C for several months *prior* to lyzation. For *in situ* lyzation, the following procedure should be followed:

1. Remove the cell culture medium.

2. Add 250 μL of the lysis solution/2-mercaptoethanol mixture for up to 5×10^6 cells or 500 μL for 5×10^6–10^7 cells to the cell culture dish.

3. Rock the culture dish while tapping the side for a few seconds to completely cover the cells with the mixture. Let the mixture react for 1–2 min.

4. Continue with **step 4** from the procedure for pelleted cells.

For pelleted cells, vortex the pellet to loosen cells and continue as follows:

1. Add 250 μL of lysis solution/2-mercaptoethanol mixture for up to 5×10^6 cells or 500 μL for 5×10^6–10^7 cells.

2. Vortex or pipette thoroughly until all clumps disappear.

3. Pipette the lysed cells into a GenElute™ filtration column, a blue insert with a 2.0 mL receiving tube (*see* **Note 4**).

4. Centrifuge at 12,000–16,000 × g (*see* **Note 5**) for 2 min.

5. Discard the filtration column.

6. Add an equal volume of 70 % ethanol solution, 250 or 500 μL, to the filtered lysate.

7. Vortex or pipette thoroughly to mix.

8. Pipette up to 700 μL of lysate/ethanol mixture into a GenElute™ binding column, a colorless insert with a red o-ring seated in a 2.0 mL receiving tube. If the volume of lysate/ethanol mixture exceeds 700 μL, the RNA must be bound to the column in two steps.

9. Centrifuge at 12,000–16,000 × g for 15 s.

10. Retain the collection tube and discard the flow-through liquid.

11. Repeat **steps 9–10** if any remaining lysate/ethanol mixture is left.

12. Continue the procedure using the On-Column DNase I Digestion Set by pipetting 250 μL of wash solution 1 into the binding column and centrifuge at 12,000–16,000 × g for 15 s.

13. Add 80 μL of the DNase I/digest buffer mixture directly onto the filter in the binding column.

14. Incubate at room temperature for 15 min.

15. Pipette 250 μL of wash solution 1 into the binding column and centrifuge at 12,000–16,000 × g for 15 s.

16. Transfer the binding column into a fresh 2.0 mL collection tube.

17. Pipette 500 μL of wash solution 1 into a binding column of the GenElute™ Mammalian Total RNA Miniprep Kit.

18. Centrifuge at 12,000–16,000 × g for 15 s.

19. Transfer the binding column into a fresh 2.0 mL collection tube.

20. Discard the flow-through liquid and the original collection tube.

21. Pipette 500 μL of the ethanol-containing wash solution 2 into the column.

22. Centrifuge at 12,000–16,000 × g for 15 s.

23. Retain the collection tube and discard the flow-through liquid.

24. Pipette 500 μL of wash solution 2 into the column.

25. Centrifuge at 12,000–16,000 × g for 2 min. If any residual wash solution 2 is seen on the surface of the binding column, centrifuge the column for an additional 1 min at 12,000–16,000 × g. Empty and reuse the collection tube if an additional centrifugation step is needed.

26. Transfer the binding column to a fresh 2.0 mL collection tube.

27. Pipette 50 μL of elution solution into the binding column.

28. Centrifuge at 12,000–16,000 $\times g$ for 1 min.

29. If more than 50 μg of RNA is expected, repeat **steps 27** and **28**, collecting both eluates in the same tube.

30. Purified RNA is ready for immediate use or store at –80 °C for several months.

3.2.2   Liver Tissue

To yield intact RNA, the liver tissue must be harvested as quickly as possible. Tissue may be instantly flash-frozen in liquid nitrogen and stored at –80 °C for several months *prior* to RNA extraction or directly harvested after collection. Alternatively, the tissue may be kept in RNA*later®* stabilization solution and stored for 1 day at 37 °C, 1 week at 25 °C, 1 month at 4 °C, or at –20 °C or colder temperatures for longer-term storage.

1. Quickly slice and weigh of a piece of fresh, frozen, or RNA*later®* stabilized liver tissue for up to 40 mg per preparation. Do not allow frozen tissue to thaw before disruption.

2. Add 500 μL of lysis solution/2-mercaptoethanol mixture (*see* **Note 6**).

3. Homogenize immediately until no visible pieces remain using a rotor-stator homogenizer or a disposable RNase-free pellet mixer.

4. Pipette the homogenized tissue into a GenElute™ filtration column, a blue insert with a 2.0 mL receiving tube.

5. Continue procedure from **step 5** in Subheading 3.2.1.

3.3   RNA-DNA Quantification and Purity Control

Quantification of RNA is recommended, as the same amounts of RNA should be used when comparing different samples. Several quantification methods are routinely applied, including spectrophotometric assays (Nanodrop®, Thermo Scientific, USA), capillary gel electrophoresis (QIAxcel®, Qiagen, USA), microfluidic analysis (Experion™, Bio-Rad, USA), or fluorescent dye detection systems (RiboGreen®, Applied Biosystems, USA). In this protocol, a spectrophotometric assay is used.

1. Open the sampling arm of the NanoDrop® ND-1000 spectrophotometer and pipette 2 μL of purified RNA or blank control, elution solution, on the lower pedestal (*see* **Note 7**).

2. Close the sampling arm and measure the absorbance at 260 nm (A_{260}) and 280 nm (A_{280}) (*see* **Note 8**).

Optionally, the integrity of the RNA can be checked by gel electrophoresis and detection of ribosomal RNA (rRNA), which makes up >80 % of total RNA in mammalian cells. The intact total RNA run on a denaturing gel will have sharp 28S rRNA and 18S rRNA bands. rRNA forms two sharp bands in the gel. The 28S rRNA band should be approximately twice as intense as the 18S rRNA band. Alternatively, 2100 Bioanalyser (Agilent Technologies, USA) is an easy-to-use device to assess RNA integrity of 12 samples for 1 biochip. Its algorithm gives a RNA Integrity Number (RIN), which scales from 0 to 10, where 10 is the highest quality. The RIN must be between 8 and 10 for further RT-qPCR analysis.

3.4   cDNA Synthesis

It is recommended that the reverse transcription step be carried out in duplicate or triplicate.

1. Thaw the 5× iScript™ reaction mix and the iScript™ reverse transcriptase at room temperature.

2. Prepare the complete reaction mix in a total volume of 40 μL (Table 2).

3. Incubate complete reaction mix (Table 3) in the iCycler iQ™.

3.5   cDNA Purification

Although purification of cDNA is optional, it is highly recommended to include this step. It is designed for rapid purification of single-stranded or double-stranded PCR amplification products (i.e., 100 base pairs to 10 kilobase) from the other components in the reactions, such as excess primers, nucleotides, DNA polymerase, oil and salts. Different cDNA purification kits are commercially available (e.g., QIAquick PCR Purification Kit® from Qiagen,

Table 2
Reaction mix for cDNA synthesis. 'X' represents the sample volume to obtain the required RNA amount

Reagent	Volume per reaction (μL)
5× iScript™ reaction mix	8
iScript™ reverse transcriptase	2
Nuclease-free water	$30 - x$
RNA template (200 fg to 2 μg total RNA)	x

Table 3
Reaction settings for cDNA synthesis

Time (min)	Temperature (°C)
5	25
30	42
5	85
Hold (optional)	4

DNAclear™ kit from Applied Biosystems, Jetquick PCR Purification Kit from Genomed). The protocol here below is outlined using the GenElute™ PCR Clean-Up Kit (Sigma, USA).

1. Insert a GenElute™ plasmid mini spin column, with a blue o-ring, into a provided collection tube if not already assembled.

2. Add 0.5 mL of column preparation solution to each GenElute™ plasmid mini spin column.

3. Centrifuge at $12,000 \times g$ for 30 s to 1 min.

4. Discard the eluate.

5. Add 5 volumes of binding solution to 1 volume of the PCR reaction and mix (i.e., add 500 μL of binding solution to 100 μL of the PCR reaction).

6. Transfer the solution into the binding column.

7. Centrifuge the column at $12,000–16,000 \times g$ for 1 min.

8. Retain the collection tube and discard the eluate.

9. Replace the binding column into the collection tube.

10. Apply 0.5 mL of diluted ethanol-containing wash solution to the column.

11. Centrifuge at $12,000–16,000 \times g$ for 1 min.

12. Retain the collection tube and discard the eluate.

13. Replace the binding column into the collection tube.

14. Centrifuge at $12,000–16,000 \times g$ for 2 min without any additional wash solution.

15. Discard any residual eluate as well as the collection tube.

16. Transfer the binding column to a fresh 2.0 mL collection tube.

17. Apply 50 μL of elution solution or water (*see* **Note 9**) to the center of each column.

18. Incubate at room temperature for 1 min.

19. Centrifuge the binding column at $12,000–16,000 \times g$ for 1 min.

20. The PCR amplification product is now present in the eluate and is ready for immediate use or storage at –20 °C.

3.6 Real-Time qPCR   The first step of the real-time qPCR process is to gather information about the DNA sequence of the target gene (i.e., Cx26, Cx32, and Cx43) to be used for primer design (*see* **Note 10**). The real-time qPCR procedure, described in this chapter, is optimized for primers targeting Cx26, Cx32, and Cx43 (Table 1). Next to the target genes, appropriate reference genes should be selected as internal controls for normalization purposes. Reference gene RNAs should be stably expressed in all samples of the study and their abundances should show strong correlation with the total amounts of RNA present in the samples [19]. Normalization against a single reference gene is not acceptable unless clear evidence is presented confirming its invariant expression under the experimental conditions described. Ideally, a pool of candidate reference genes (Table 1) should be used and the optimal number and choice of the reference genes must be experimentally determined during data analysis (*see* Subheading 3.6).

It is also strongly advised to use non-template controls (NTC), thus reaction mixes containing no DNA template as negative control (*see* **Note 11**), on each plate or batch of samples. The RT-qPCR system applied in this protocol is the StepOnePlus™ real-time PCR system (Applied Biosystems, USA) combined with the TaqMan® reagents (Applied Biosystems, USA). The TaqMan® reagents consist of two primers and a hydrolysis probe. The primers are designed to amplify the target, while the hydrolysis probe is designed to hybridize to the target and generate fluorescence when the target is amplified (Fig. 1). Alternatively, SYBR® Green reagents (Life Technologies, USA) can be used (*see* **Note 12**).

1. Prepare the DNA standard 1 by pooling cDNA from the samples. A minimum of 7.5 μL cDNA should be pooled per target and reference gene if the standards will be analyzed in triplicate (i.e., prepare 67.5 μL when applying three target genes and six reference genes). Every sample should provide an equal volume for the preparation of DNA standard 1 (i.e., if ten samples should be analyzed, 6.75 μL cDNA of every sample should be pooled to obtain standard 1).

2. Vortex thoroughly.

3. Prepare standard 2 by diluting standard 1 with a dilution factor 5 with nuclease-free water.

4. Repeat **steps 2** and **3** to obtain a serial dilution of minimum 5 standards.

5. Prepare 20 μL reaction volume for each standard, sample, and NTC (Table 4).

6. Each standard, NTC, and sample is run in triplicate (i.e., 3 wells per sample). Therefore, the sample reaction mix volumes should be tripled.

7. Seal the 96-well plate and centrifuge at $1600 \times g$ for 1 min at 4 °C (*see* **Note 13**).

8. Samples are placed in a MicroAmp® optical 96-well reaction plate and assay run in the StepOnePlus™ machine as instructed in the manufacturer's guidelines (*see* **Note 14**).

9. Select the appropriate ramp speed for the instrument run (Table 5).

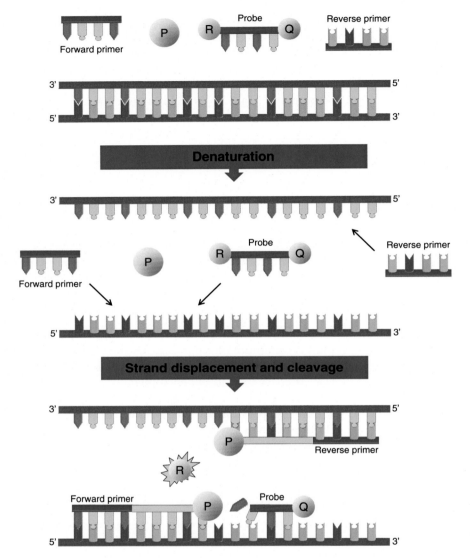

Fig. 1 Mechanism of real-time detection by hydrolysis probes. The reaction mix of real-time qPCR consists of cDNA, forward and reverse primer, DNA polymerase enzyme (P), and hydrolysis probe. The probe contains a reporter dye (R) at the 5′ end and a quencher dye (Q) at the 3′ end. When the probe is intact, the proximity of the reporter dye to the quencher dye results in suppression of the reporter fluorescence. Upon denaturation and polymerization, the DNA polymerase enzyme cleaves the hydrolysis probe between the reporter and the quencher only if the probe hybridizes to the target, resulting in increased fluorescence of the reporter. The polymerization of the strand is not harmed and continues

10. The associated StepOnePlus™ software will automatically calculate the quantification cycle (C_q) (*see* **Note 15**). C_q values exceeding 40 are questionable because they imply low efficiency (*see* **Note 16**) and generally should not be reported. However, the use of such arbitrary C_q cut-offs is not ideal, because they may be either too low, eliminating valid results, or too high, increasing false positive results [19].

Table 4
Reaction mix for samples and standards in real-time qPCR

Reagent	Volume (μL)
Taqman® Universal PCR Master Mix (2×)	10
Assay-on-Demand™ Gene Expression Assay Mix (20×)	1
cDNA/Standards	2
Nuclease-free water	7

Table 5
Run settings of real-time qPCR. The denaturation and anneal and extension step run for 40 cycles

Function	Time	Temperature (°C)
Incubation	2 min	50
DNA polymerase activation	20 s	95
Denaturation	1 s	95
Anneal/extension	20 s	60

3.7 Data Processing

Many different methods have been proposed to normalize qPCR data [22]. The use of reference genes is undoubtedly the most popular and adequate approach. However, accurate normalization can only be performed after validation of the candidate reference genes. Normalization against adequate reference genes corrects for variable sample mass, nucleic acid extraction efficiency, reverse transcription efficiency, and pipette calibration errors. Different software systems can be used for this purpose, including geNorm [23], NormFinder [24], and BestKeeper [25]. These algorithms determine the most stable reference genes from a set of tested genes in a given sample panel. From this, a gene expression normalization factor can be calculated for each tissue sample based on the geometric mean of a user-defined number of reference genes. GeNorm calculates the gene expression stability measure M for a reference gene as the average pairwise variation V for that gene with all other tested candidate reference genes. Stepwise exclusion of the gene with the highest M value allows ranking of the tested genes according to their expression stability [23, 26]. The geNorm is currently integrated in the qbase+ software, which additionally to the first version gives a fully automatic and expert result report, handles missing data, and identifies single best reference genes (Fig. 2). Besides geNorm, the qbase+ software contains different normalization methods to accommodate a wide range of experiments, post-PCR quality control, interrun calibration, and statistical analysis wizard. The qbase+ helps applying MIQE compliant

---

Data

Input

geNorm analysis was initiated on 35 samples and 8 reference targets.

Missing values

Experiment design

The run lay-out is perfect. All samples are measured in the same run for a given reference target (i.e. sample maximization strategy according to Hellemans et al., Genome Biology, 2007).

Results

Optimal reference target selection

The optimal number of reference targets in this experimental situation is 3 (geNorm V < 0.15 when comparing a normalization factor based on the 3 or 4 most stable targets). As such, the optimal normalization factor can be calculated as the geometric mean of reference targets Gapdh, Hmbs and B2m.

Reference target stability

Medium reference target stability (average geNorm M ≤ 1.0). This is typically seen when evaluating candidate reference targets on a heterogeneous set of samples (e.g. treated cultured cells, cancer biopsies, or samples from different tissues). More reference values in Table 1 in Hellemans et al., Genome Biology, 2007.

Fig. 2 GeNorm result report using the qbase$^+$ software. Example of a fully automatic and expert result report with handling of missing data and identification of single best reference genes

procedures and guides the experimenter to highest quality results. Unlike geNorm, NormFinder adopts a model-based approach to give a score to the two most stable reference genes with the least intragroup and intergroup variation. Stability is expressed as a value in arbitrary units. Furthermore, NormFinder possesses the ability to discriminate between sample variability and bias among several groups [24]. BestKeeper determines the variability in expression of a set of reference genes by analyzing Cq values and classifying variability by the coefficient of variance and the standard deviation. To define the most stable reference gene, the software generates an index, which finally is compared to each candidate reference gene. This comparison results in a value for the Pearson correlation coefficient and probability, which are then allocated to each candidate reference gene [25]. Using this software system, relative alterations (i.e., fold change) in RNA levels can be calculated according to the Livak $2^{-\Delta\Delta Cq}$ formula [27]. Using the latter, data are presented as the fold change in gene expression normalized to the selected reference genes and relative to the untreated control.

4   Notes

1. Reagents from the GenElute™ Mammalian Total RNA Miniprep Kit and the 5× iScript™ reaction mix may contain some precipitation upon thawing. It should be mixed thoroughly to resuspend the precipitation.

2. Lysis solution and wash solution 1 of the GenElute™ Mammalian Total RNA Miniprep Kit contain guanidine

thiocyanate, which is a potent chaotripic agent and irritant. Wear gloves, safety glasses, and suitable protective clothing when handling these solutions or any reagent provided with the kit.

3. In general, RT-qPCR analysis requires working with small volumes and tubes, including handling volumes of less than 1 μL. This necessitates correct standard operation procedures for pipetting small volumes. If the pipette tip is submerged into the sample or reagent, the delivered volume to the reaction will be larger than intended. Instead, the pipette tip should be touched to the surface of the sample or reagent (Fig. 3). To ensure reproducibility, pipettors should be calibrated on a regular basis (i.e., at least every 6 months). Alternatively, automated liquid handling systems (e.g., epMotion® 5070, Eppendorf, Germany) can be used for pipetting tasks in the RT-qPCR procedure. These automated systems help to eliminate manual pipetting errors and maximize the reproducibility.

4. The filtration step removes cellular debris and shears DNA. This step may be omitted with fewer than 1×10^6 cultured cells.

Fig. 3 Correct pipetting procedure. (**a**) Avoid submerging the pipette tip into the sample or reagent. The delivered volume to the reaction will be larger than intended. (**b**) Instead, the pipette tip should be touched to the surface of the sample

5. Centrifugation speeds are indicated in units of g. Convert the relative centrifugal force (RCF) to rotation per min (rpm) according to the following formula: $RCF = 1.118 \times 10^{-5} \times radius$ (in cm) $\times rpm^2$.

6. For larger amounts, scale up the volume of the lysis solution/2-mercaptoethanol mixture proportionally. Divide the lysate into 500–700 μL aliquots and process through separate filtration and binding columns.

7. After each sample measurement the liquid on the upper and lower pedestals should be removed with a soft laboratory wipe to prevent sample carryover in successive measurements. After measuring a large number of samples, it is recommended to clean the pedestals thoroughly using 2 μL water aliquots.

8. An A_{260} reading of 1.0 is equivalent to about 40 μg/mL of RNA. The A_{260}/A_{280} ratio provides an indication of the RNA purity, in particular the presence of DNA or residual phenol. Pure RNA has an A_{260}/A_{280} ratio between 1.8 and 2.1.

9. When eluting with water, make sure that the pH of the water is between 5.5 and 8.5. Elution may also be performed using the elution solution diluted tenfold with water.

10. The most convenient way to gather information about the DNA sequence of the target gene is by consulting the GenBank™ database maintained at the National Center for Biotechnology Information (NCBI). Importantly, as one searches for DNA sequences for connexin genes, one should consider the correct designation. Indeed, two nomenclature systems are currently used to name the different connexin species [28]. A first system is based on the predicted molecular weight in kilodaltons (i.e., Cx26 refers to a connexin with a molecular weight predicted by cDNA sequencing of 26 kDa). A second system divides the connexin family into four subclasses according to genetic origin, overall sequence similarity, and length of cytoplasmic domain. This classification system forms the basis for the current official nomenclature of connexin genes. Hence, the corresponding gene names for Cx26, Cx32, and Cx43 are GJB2, GJB1, and GJA1, respectively, for human, and Gjb2, Gjb1, and Gja1 for other species. The information about the DNA sequence can be used to carefully choose the oligonucleotide primers, which can be performed using software, such as Oligo® [29], Primer3 [30], and PRIMO [31].

11. Amplification in the NTC may indicate genomic contamination of samples or amplicon cross-contamination. RNA samples that contain DNA should be treated with DNase. When possible, probes should be designed spanning an exon-exon junction to avoid amplification of genomic DNA. It is essential to take precautions in order to clear samples from RNases and DNases. In case of contamination, all reagents (e.g., master mix) and stock

buffers should be replaced, and all PCR areas should be cleaned meticulously.

12. Hydrolysis Taqman® reagents have the advantage to contain an increased specificity compared to traditional primer systems and to provide capability for use in multiplex systems. A limitation is that the system requires synthesis of a unique fluorogenic probe. The SYBR® Green reagents use a double-stranded DNA binding dye to detect PCR products, as they accumulate during PCR cycles. This system is financially more favorable and allows melt curve analysis. However, the dye binds nonspecifically to all double-stranded DNA sequences. This could yield false positive results. To avoid false positive signals, formation of nonspecific products should be identified using melt curve or gel analysis.

13. The reaction mix should be at the bottom of each well of the reaction plate. If not, centrifuge the reaction plate again for a longer period of time at a higher centrifugation speed. It is important not to allow the bottom of the reaction plate to become smudged. Fluids and other impurities that adhere to the bottom of the reaction plate can cause a contamination and create an abnormally high background signal.

14. The instrument needs monthly maintenance, including calibration. In addition, make sure that the instrument is set on the correct dye and filter settings.

15. The nomenclature describing the fractional PCR cycle used for quantification is inconsistent, with threshold cycle (C_t), crossing point (C_p), and take-off point (TPF) currently used in the literature. The MIQE guidelines therefore propose to use the term quantification cycle (C_q) [19].

16. The efficiency of each PCR assay needs to be taken into account for quality control purposes. If the efficiency is not between 90 and 110%, the PCR assay requires further optimization before results can be considered valuable and suitable for further purposes. The efficiencies for the target and reference genes should be approximately equal.

Acknowledgements

This work was financially supported by the grants of Agency for Innovation by Science and Technology in Flanders (IWT grant 131003), the University Hospital of the Vrije Universiteit Brussel-Belgium (Willy Gepts Fonds UZ-VUB), the Fund for Scientific Research-Flanders (FWO grants G009514N and G010214N), the European Research Council (ERC Starting Grant 335476), the University of São Paulo-Brazil, and the Foundation for Research Support of the State of São Paulo (FAPESP SPEC grant 2013/50420-6).

References

1. Oyamada M, Oyamada Y, Takamatsu T (2005) Regulation of connexin expression. Biochim Biophys Acta 1719:6–23

2. D'hondt C, Iyyathurai J, Vinken M et al (2013) Regulation of connexin- and pannexin-based channels by post-translational modifications. Biol Cell 105:373–398

3. Piechocki MP, Toti RM, Fernstrom MJ et al (2000) Liver cell-specific transcriptional regulation of connexin32. Biochim Biophys Acta 1491:107–122

4. Koffler LD, Fernstrom MJ, Akiyama TE et al (2002) Positive regulation of connexin32 transcription by hepatocyte nuclear factor-1alpha. Arch Biochem Biophys 407:160–167

5. Vinken M, De Rop E, Decrock E et al (2009) Epigenetic regulation of gap junctional intercellular communication: more than a way to keep cells quiet? Biochim Biophys Acta 1795:53–61

6. Oyamada M, Takebe K, Oyamada Y (2013) Regulation of connexin expression by transcription factors and epigenetic mechanisms. Biochim Biophys Acta 1828:118–131

7. Alwine JC, Kemp DJ, Stark GR (1977) Method for detection of specific RNAs in agarose gels by transfer to diazobenzyloxymethyl-paper and hybridization with DNA probes. Proc Natl Acad Sci U S A 74:5350–5354

8. Parker RM, Barnes NM (1999) mRNA: detection by in situ and northern hybridization. Methods Mol Biol 106:247–283

9. Hod Y (1992) A simplified ribonuclease protection assay. Biotechniques 13:852–854

10. Saccomanno CF, Bordonaro M, Chen JS et al (1992) A faster ribonuclease protection assay. Biotechniques 13:846–850

11. Weis JH, Tan SS, Martin BK et al (1992) Detection of rare mRNAs via quantitative RT-PCR. Trends Genet 8:263–264

12. Wang T, Brown MJ (1999) mRNA quantification by real time TaqMan polymerase chain reaction: validation and comparison with RNase protection. Anal Biochem 269:198–201

13. Derveaux S, Vandesompele J, Hellemans J (2010) How to do successful gene expression analysis using real-time PCR. Methods 50:227–230

14. Loewenstein WR, Kanno Y (1967) Intercellular communication and tissue growth. I. Cancerous growth. J Cell Biol 33:225–234

15. Revel JP, Karnovsky MJ (1967) Hexagonal array of subunits in intercellular junctions of the mouse heart and liver. J Cell Biol 33:C7–C12

16. Maes M, Cogliati B, Crespo YS et al (2015) Roles of connexins and pannexins in digestive homeostasis. Cell Mol Life Sci 72:2809–2821

17. Vinken M, Henkens T, De Rop E et al (2008) Biology and pathobiology of gap junctional channels in hepatocytes. Hepatology 47:1077–1088

18. Maes M, Crespo YS, Willebrords J et al (2015) Connexin and pannexin signaling in gastrointestinal and liver disease. Transl Res 166:332–343

19. Bustin SA, Benes V, Garson JA et al (2009) The MIQE guidelines: minimum information for publication of quantitative real-time PCR experiments. Clin Chem 55:611–622

20. Bustin SA, Beaulieu JF, Huggett J et al (2010) MIQE précis: practical implementation of minimum standard guidelines for fluorescence-based quantitative real-time PCR experiments. BMC Mol Biol 11:74

21. Micke P, Ohshima M, Tahmasebpoor S et al (2006) Biobanking of fresh frozen tissue: RNA is stable in nonfixed surgical specimens. Lab Invest 86:202–211

22. Huggett J, Dheda K, Bustin S et al (2005) Real-time RT-PCR normalisation; strategies and considerations. Genes Immun 6:279–284

23. Vandesompele J, De Preter K, Pattyn F et al (2002) Accurate normalization of real-time quantitative RT-PCR data by geometric averaging of multiple internal control genes. Genome Biol 3 RESEARCH0034

24. Andersen CL, Jensen JL, Orntoft TF (2004) Normalization of real-time quantitative reverse transcription-PCR data: a model-based variance estimation approach to identify genes suited for normalization, applied to bladder and colon cancer data sets. Cancer Res 64:5245–5250

25. Pfaffl MW, Tichopad A, Prgomet C et al (2004) Determination of stable housekeeping genes, differentially regulated target genes and sample integrity: BestKeeper - Excel-based tool using pair-wise correlations. Biotechnol Lett 26:509–515

26. Hellemans J, Vandesompele J (2014) Selection of reliable reference genes for RT-qPCR analysis. Methods Mol Biol 1160:19–26

27. Livak KJ, Schmittgen TD (2001) Analysis of relative gene expression data using real-time quantitative PCR and the $2^{-\Delta\Delta Ct}$ method. Methods 25:402–408

28. Sohl G, Willecke K (2004) Gap junctions and the connexin protein family. Cardiovasc Res 62:228–232

29. Rychlik W, Rhoads RE (1989) A computer program for choosing optimal oligonucleotides for filter hybridization, sequencing and in vitro amplification of DNA. Nucleic Acids Res 17:8543–8551

30. Rozen S, Skaletsky H (2000) Primer3 on the WWW for general users and for biologist programmers. Methods Mol Biol 132:365–386

31. Li P, Kupfer KC, Davies CJ et al (1997) PRIMO: a primer design program that applies base quality statistics for automated large-scale DNA sequencing. Genomics 40:476–485

Chapter 2

DNA Methylation Analysis of Human Tissue-Specific Connexin Genes

Xiaoming Liu and Canxia Xu

Abstract

Connexins are the structural proteins of gap junctions and their functioning as tumor suppressors is well known. Epigenetic modifications, such as methylation of connexin genes, play important roles in regulating gene expression. Over the past decade, several methods have been applied to characterize DNA methylation-specific loci of connexin genes. This chapter describes analysis of selective connexin32 and connexin43 gene DNA methylation in human gastric tissues using methylation-specific PCR, bisulfite-specific PCR sequencing as well as MassArray techniques.

Key words Connexin, Gap junction, Methylation-specific PCR, Bisulfite-specific PCR sequencing, MassArray, DNA methylation, Normal gastric mucosa, Non-atrophic gastritis, Chronic atrophic gastritis, Intestinal metaplasia, Dysplasia, Gastric carcinoma

1 Introduction

Gap junctions are cellular channels necessary to coordinate cell function by allowing adjacent cells to directly share ions and molecules less than 1 kDa [1]. The connexin (Cx) family is a multigene group of gap junction proteins. Gap junctional intercellular communication (GJIC) directly potentiates cells to cooperate electrically or metabolically. In both physiological and pathological conditions, connexin functions are not always linked to their roles in GJIC. For instance, the cytoplasmic carboxy tails of Cx32, also known as gap junction protein β1 or GJB1, and Cx43, also called gap junction protein α1 or GJA1, interact with the PDZ domain of zona occludens protein 1, linking them to the cytoskeleton [2, 3]. Interaction of the Cx43 cytoplasmic carboxy tail with β-catenin may influence Wnt signaling and can reside in the nucleus, resulting in inhibition of cell growth via interfering with key steps of cell cycle regulation [4–6]. Reduced connexin expression and loss of GJIC release initiated cells from growth control exerted by surrounding normal cells and thus allow their clonal expansion,

Mathieu Vinken and Scott R. Johnstone (eds.), *Gap Junction Protocols*, Methods in Molecular Biology, vol. 1437,
DOI 10.1007/978-1-4939-3664-9_2, © Springer Science+Business Media New York 2016

which can be considered as early events in cancer development. Moreover, Cx26, Cx32, and Cx43 have been reported to act as tumor suppressors in human renal cell carcinoma (RCC) and breast cancer. Downregulation of connexin expression in precancerous lesions or cancer tissues may occur at the transcriptional level, of which a prominent mechanism is cancer-specific hypermethylation of their gene promoters [7].

Connexins display high turnover rates with half-lives of 1.5–5 h [8]. In the last few years, it has become clear that epigenetic processes are essentially involved in connexin gene transcription. The most extensively studied epigenetic mechanism is DNA methylation, which occurs almost exclusively at cytosine residues in CpG dinucleotides [9]. Higher frequencies of CpG dinucleotides are found in so-called CpG islands that comprise 1–2 % of the genome [10]. CpG islands are typically present in gene promoter regions and are generally unmethylated. It has been well documented that methylation of CpG islands in connexin gene promoter regions is associated with the transcriptional silencing [11, 12]. DNA hypermethylation contributes to oncogenesis by point mutation and inactivation of tumor suppressor genes. Methylated connexin genes are known to underlie various diseases and accumulation of methylated CpG dinucleotides in connexin gene promoters is frequently observed in neoplastic cells [3, 13, 14]. Not surprisingly, the DNA demethylating agent 5-aza-2′-deoxycytidine suppresses growth of RCC tumors in a xenograft model and breast cancer cell lines by restoring Cx32 and Cx26 expression [5] as well as GJIC [15], indicating the reversed tumorigenicity by re-introduction of functional copies into cells lines using demethylation therapy. Based on the predictable and therapeutic role of DNA methylation in human disease, attention has been paid to this epigenetic modification in gene promoter regions as a gatekeeper of connexin expression [16]. In the present chapter, three methods are described for analyzing the methylation status of connexin CpG islands in normal gastric mucosa as well as tissues in pathological conditions, including non-atrophic gastritis, chronic atrophic gastritis, intestinal metaplasia, dysplasia, and gastric carcinoma.

2   Materials

2.1   Genomic DNA Extraction from Tissues

1. Wizard® Genomic DNA purification kit (Promega, USA).

2. 15 mL centrifuge tubes.

3. Small homogenizer.

4. Liquid nitrogen for animal tissue grinding (eventually replacing the small homogenizer).

5. Mortar and pestle for animal tissue grinding (eventually replacing the small homogenizer).

6. Isopropanol at room temperature.

7. 70 % ethanol at room temperature.

8. Water bath at 65 °C.

9. Water bath at 37 °C.

2.2 Sodium Bisulfite Conversion of Unmethylated Cytosines in DNA

EpiTect Bisulfite kit (Qiagen, USA).

2.3 PCR Reagents

1. 10× PCR buffer (Qiagen, USA).

2. 10 mM dNTPs (Amersham Pharmacia Biotech Products Inc., USA).

3. 10 mM forward primer.

4. 10 mM reverse primer.

5. HotStarTaq DNA polymerase (Qiagen, USA).

6. Bisulfite-treated genomic DNA.

7. DNAse-free and RNAse-free water.

2.4 Methylation-Specific PCR

GelDoc XR system (Bio-Rad, USA).

2.5 Bisulfite-Specific PCR Sequencing

1. BigDye® XTerminator™ purification kit (Applied Biosystems, USA): XTerminator solution and SAM™ solution.

2. BigDye® Terminator v1.1 cycle sequencing kit (Applied Biosystems, USA): M-13 tailed sequencing primers.

3. QIAprep® Miniprep kits (Qiagen, USA).

4. QIAquick gel extraction kit (Qiagen, USA).

2.6 MassArray Assay

1. MassARRAY® system (Sequenom Inc., USA).

2. MassCLEAVE kit (Sequenom Inc., USA).

3 Methods

As such, three methods, namely methylation-specific PCR (MSP), bisulfite-specific PCR (BSP) sequencing, and MassArray analysis, were selected to detect Cx32 and Cx43 gene promoter methylation (Table 1). These three approaches are based on the same principle of genomic DNA extraction and bisulfite-modified conversion according to the steps as described below (Fig. 1).

Table 1
Most common approaches for analyzing DNA methylation of selected gene sequences

Method	Annotation
MSP	Methylation-specific polymerase chain reaction
BSP	Bisulfite sequencing polymerase chain reaction of clones
qPCR	Subsequently after restriction
Bisulfite pyrosequencing	
COBRA	Combined bisulfite restriction analysis
Ms-SNuPE	Methylation-sensitive single nucleotide primer extension
MethyLight	Fluorescent-based real-time polymerase chain reaction (TaqMan®) in combination with bisulfite treatment
MassArray	
Compare-MS	Compare-methylation sequencing
AP-PCR	Arbitrarily primed polymerase chain reaction
MSRE-PCR	Methylation-sensitivity restriction enzymes combined with polymerase chain reaction

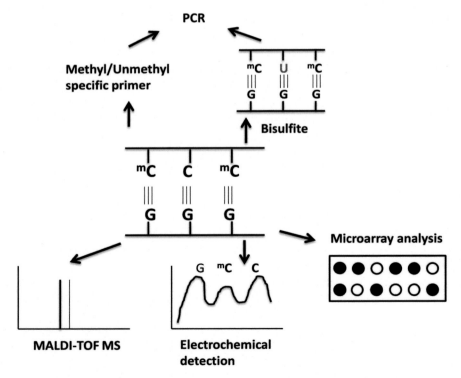

Fig. 1 Overview of DNA methylation detection techniques. (MALDI-TOF MS, matrix-assisted laser desorption/ionization time-of-flight mass)

3.1 Genomic DNA Extraction from Human Gastric Tissues

1. Grind tissue in liquid nitrogen using a pre-chilled mortar and pestle.

2. Allow the liquid nitrogen to evaporate, add 10–20 mg fresh tissue to 600 µL of chilled nuclei lysis solution in a 1.5 mL microcentrifuge tube, and homogenize for 10 s (*see* **Note 1**).

3. Add 3 µL of RNase solution to the cell or animal tissue nuclei lysate and mix. Incubate for 15–30 min at 37 °C and cool to room temperature.

4. Add 200 µL of protein precipitation solution, mix vigorously at high speed for 20 s, and chill the sample on ice for 5 min.

5. Centrifuge at 13,000–16,000×*g* for 4 min. The precipitated protein will form a tight white pellet.

6. Carefully transfer the supernatant containing DNA, leaving the protein pellet behind, to a new tube and add 600 µL isopropanol (*see* **Note 2**).

7. Mix gently by inversion until the white thread-like strands of DNA form a visible mass and centrifuge at 13,000–16,000×*g* for 1 min.

8. Carefully decant supernatant and add 600 µL 70 % ethanol. Gently invert the tube several times to wash the DNA and centrifuge at 13,000–16,000×*g* for 1 min.

9. Carefully vacuum-aspirate the ethanol and air-dry the pellet for 15 min (*see* **Note 3**).

10. Redissolve the DNA in 100 µL of DNA rehydration solution overnight at 4 °C or for 1 h at 65 °C. Mix the solution periodically by gently tapping the tube.

11. Store the DNA at 2–8 °C.

3.2 Sodium Bisulfite Conversion of Unmethylated Cytosine

Bisulfite reaction-based assays rely on the chemical conversion of cytosine to uracil. Sodium bisulfite rapidly deaminates position 5 methylated cytosine (5mC) residues to uracil compared to the slower deamination of 5mC to thymine [15, 16]. Bisulfite-based techniques allow for the analysis of this non-CpG methylation, while other techniques cannot, making this one of the major advantages of using bisulfite reaction-based methods.

3.2.1 Bisulfite DNA Conversion

1. Thaw DNA and dissolve the required number of aliquots of bisulfite mix by adding 800 µL RNase-free water to each aliquot. Mix until completely dissolved, which may take up to 5 min.

2. Prepare the bisulfite reactions in 200 µL PCR tubes up to a total volume of 140 µL:

    (a) 1 ng to 2 µg DNA solution.

    (b) RNase-free water to 20 µL total.

    (c) 85 µL bisulfite mix.

    (d) 35 µL DNA protect buffer.

3. Close the PCR tubes and mix the bisulfite reactions thoroughly. Store the tubes at room temperature (*see* **Note 4**).

4. Set a thermal cycler with the following bisulfite conversion conditions: denaturation 5 min at 99 °C, incubation 25 min at 60 °C, denaturation 5 min at 99 °C, incubation 85 min at 60 °C, denaturation 5 min at 99 °C, incubation 175 min at 60 °C and hold up to overnight at 20 °C.

5. Split each tube into three different tubes and start the PCR program.

3.2.2 Clean-Up of Bisulfite-Converted DNA

1. Briefly centrifuge the PCR tubes containing bisulfite reactions and transfer the complete bisulfite reactions to new 1.5 mL microcentrifuge tubes.

2. Add 560 µL freshly prepared buffer BL containing 10 µg/mL carrier RNA. Mix solution using a vortex and centrifuge briefly (*see* **Note 5**).

3. Place an EpiTect spin column and collection tube in a suitable rack. Transfer the mixture into EpiTect spin columns.

4. Centrifuge columns at $16,000 \times g$ speed for 1 min. Discard the flow-through and place the spin columns back in the collection tubes.

5. Add 500 µL buffer BW to the spin column and centrifuge at $16,000 \times g$ for 1 min. Discard the flow-through and place spin columns back into the collection tubes.

6. Add 500 µL buffer BD (i.e., desulfonation buffer) to the spin columns and incubate for 15 min at room temperature (*see* **Note 6**).

7. Centrifuge the columns at $16,000 \times g$ for 1 min. Discard the flow-through and place the columns back into the collection tubes.

8. Add 500 µL buffer BW and centrifuge at $16,000 \times g$ for 1 min. Discard and replace the spin columns. Repeat this step.

9. Place the spin columns into new 2 mL collection tubes and centrifuge at $16,000 \times g$ for 1 min to remove residual liquid.

10. Place the spin columns into clean 1.5 mL microcentrifuge tubes. Add 20 µL buffer EB to the center of the membrane. Elute purified DNA by centrifugation for 1 min at approximately $15,000 \times g$.

11. Store the DNA at −20 °C (*see* **Note 7**).

3.3 Primer Design and Methylation Positive Control Preparation

1. The primers for detection of Cx32 and Cx43 gene promoter methylation are designed with MethPrimer for MSP (*see* **Note 8**) and BSP (*see* **Note 9**), and EpiDesigner for MassArray analysis (Fig. 2) (Table 2).

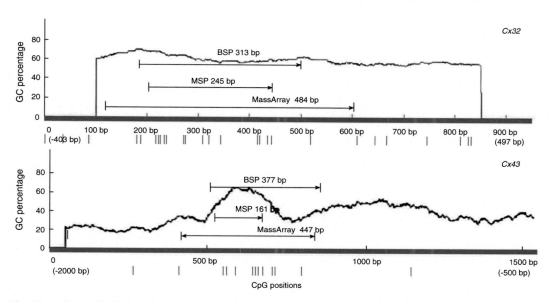

Fig. 2 Locations of primers for polymerase chain reaction for DNA methylation detection with amplicon of Cx32 and Cx43. Reprinted with permission from ref. 19

2. Given that the connexin expression is negatively associated with the gene methylation status, the primers for detection of Cx32 and Cx43 expression designed with Primer5 are required as a control for RT-PCR analysis, including the internal reference β-actin.

3.4 Connexin DNA Methylation Analysis

3.4.1 Methylation-Specific PCR

MSP is used to evaluate and quantify the connexin gene methylation status based on bisulfite reactivity. MSP uses two distinct methylation-specific primer sets for the sequence of interest. An unmethylated primer will only amplify sodium bisulfite-converted unmethylated DNA, whereas a methylated primer is specific for sodium bisulfite-treated methylated DNA [16].

1. Following bisulfite modification of tissue DNA, unmethylated and methylated reactions of MSP are carried out in a total volume of 25 μL containing 12.5 μL 2× EpiTect master mix, 2 μL 200 μM dNTP, 0.5 μL 0.4 μM of forward or reverse primer, and 10 μL RNase-free water.

2. Place the PCR tubes in the thermal cycler and start the cycling program using the following conditions: 1 cycle at 95 °C for 10 min, 38 cycles at 94 °C for 15 s, annealing temperature for 30 s, 72 °C for 30 s, and 1 cycle at 72 °C for 10 min.

3. Treat DNA of normal human peripheral blood lymphocytes with SssI methyltransferase and subject to bisulfite modification. This serves as the GC-sites methylation positive control (*see* **Note 10**). Water can be used as a negative PCR control.

Table 2
Cx32 and Cx43 primer sequences, amplified fragment size, and annealing temperature. Reprinted with permission from ref. 19

Method	Gene	Primer sequence (5′–3′)	Amplified fragment size (bp)	Annealing temperature (°C)
Real-time RT-PCR	Cx32	F: ATGAACTGGACAGGTTTGTAC	302	56
		R: ATGTGTTGCTGGTGCAGCCA		
	Cx43	F: TGCAGCAGTCTGCCTTTCGTTG	219	56
		R: CCATCAGTTTGGGCAACCTTG		
	β-actin	F: TGGACTTCGCAGCACAGCAGATGG	289	56
		R: ATCTCCTTCTGCATCCTGTCG		
MSP	Cx32 (M)	F: GGGGCGGGTGCGGCGAT	245	64
		R: CTCCGCGCCTACGTCCC		
	Cx32 (U)	F: GGGGTGGGTGTGGTGAT	245	64
		R: CTCCACACCTACATCCCAA		
	Cx43 (M)	F: AAATTGTAATATTTGGGTTTCAGCGC	156	58
		R: AATAACGCCATCTCTACTCACCG		
	Cx43 (U)	F: TTTTAAAATTGTAATATTTGGGTTTCAGTGT	161	56
		R: AATAACACCATCTCTACTCACCACA		
BSP	Cx32	F: GGTTATTTTTTTGGTGGGGTTATG	313	58
		R: ACCCAAACAAATCCCCTATAATCTC		
	Cx43	F: TGTTTTTTAAAATTGTAATATTTGGGTTTA	377	56
		R: AAAAACAAACTCATCTAACCTTCCTATTC		
MassArray	Cx32	F: CAGTTTCAGCAGTTTTTGGGTTTTTTGG	484	60
		R: TAACTCCCTATCCCCTAACTCCTTA		
	Cx43	F: ATGTTTTTGCAGGTTGGATCAGGAAAT	447	60
		R: ACCAACAAATAAAAACAAAATTATTCC		

4. Create an unmethylated DNA positive control by using the methyltransferase inhibitor 5-aza-2-deoxycytidine (*see* **Note 11**). Water can be used as a negative PCR control.

5. Detect the PCR products by gel electrophoresis using 2 % agarose gel (Fig. 3).

6. Calculate the methylation level according to the formula [M/(M + U) × 100%] using the gray values of methylation (M) and unmethylation (U) bands.

3.4.2 Bisulfite-Specific PCR Sequencing

BSP amplifies targets regardless of the gene methylation state of the internal sequence and is considered the gold standard for this type of analysis. One of the primers is fluorescently labeled at the 5′ end, so that the resulting amplicons can be identified by electrophoresis and subjected to sequencing [17]. It thus provides an inherently more accurate assessment of the gene methylation state

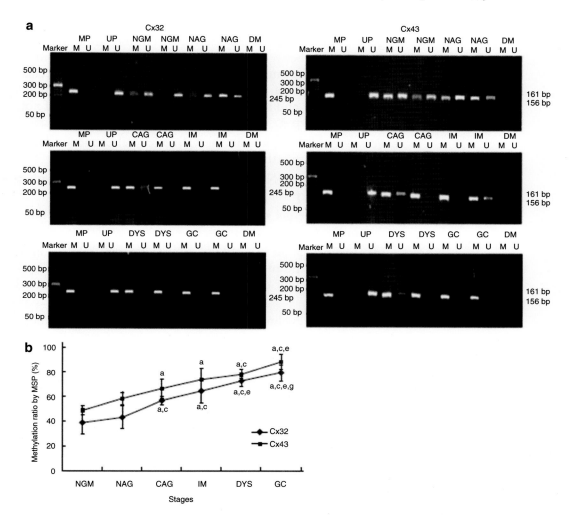

Fig. 3 Representative results for Cx32 and Cx43 gene promoters at different gastric carcinogenesis stages using methylation-specific polymerase chain reaction. (**a**) Agarose gel electrophoresis of MSP bands. (**b**) Quantified MSP bands (*CAG* chronic atrophic gastritis, *DW* diluted water as negative control, *DYS* dysplasia, *GC* gastric cancer, *IM* intestinal metaplasia, *M* methylated, *MP* methylation positive control, *NAG* non-atrophic gastritis, *NGM* normal gastric mucosa, *U* unmethylated, *UP* unmethylated positive control). Reprinted with permission from ref. 19

compared to MSP that selects for presupposed fully methylated or fully unmethylated complementary sequences.

1. The BSP PCR reaction mixture of 25 μL contains 2 μL bisulfite-modified DNA template, 12.5 μL TaKaRa Premix Taq HS, 1 μL each 10 μmol/L forward or reverse primers, and 8.5 μL deionized distilled water.

2. The annealing temperature (Ta) is set using a gradient thermal cycler (*see* **Note 12**): 1 cycle at 95 °C for 5 min, 35 cycles at 94 °C for 1 min, the target Ta for 2.5 min, 72 °C for 1 min, and 1 cycle at 72 °C for 5 min.

3. Load a fraction of the PCR amplification onto a 1.5 % agarose gel with mass ladder, allow electrophoretic migration, and view the products between 200 and 500 base pairs using standard methods. If the amplified product is of the correct molecular weight and free of nonspecific by-products, proceed to traditional cloning and sequencing as described in **step 14**. If not, proceed to **step 4**.

4. Follow the instructions in the QIAquick gel extraction kit to extract and purify the PCR bands:

   (a) Excise the DNA fragment from the agarose gel with a clean sharp scalpel.

   (b) Weigh the gel slice in a colorless tube and add 3 volumes buffer QG to 1 volume gel.

   (c) Incubate at 50 °C until the gel slice is completely dissolved and mix the tube every 2–3 min to help dissolve. Check that the color of the mixture is yellow without dissolved agarose.

   (d) Add 1 gel volume of isopropanol to the sample and mix.

   (e) Place a QIAquick spin column in a 2 mL collection tube.

   (f) Apply the sample to the QIAquick column and centrifuge for 1 min. Discard the flow-through and place the QIAquick column back in the same collection tube to bind DNA.

   (g) Add 0.75 mL buffer PE to the QIAquick column and centrifuge for 1 min. Discard the flow-through and centrifuge the QIAquick column for an additional 1 min at $12,000 \times g$.

   (h) Place the QIAquick column into a new 1.5 mL microcentrifuge tube.

   (i) Add 50 µL water to the center of the QIAquick membrane and centrifuge the column for 1 min at $16,000 \times g$ to elute DNA.

5. Set up ligation reactions for amplicon cloning by mixing 3 µL of PCR purified product, 1 µL plasmid, 1 µL T4 ligase, and 2.5 µL reaction buffer. Mix the reactions by pipetting. Incubate the reactions for 1 h at room temperature.

6. Transform normal cells as follows:

   (a) Carefully transfer 50 µL cell suspension into each prepared tube.

   (b) Gently flick the tubes to mix and place them on ice for 20 min.

   (c) Heat-shock the cells for 45–50 s in a water bath at exactly 42 °C and do not shake.

   (d) Immediately return the tubes to ice for 2 min.

7. Add 950 μL SOC medium at room temperature to the tubes containing cells transformed with ligation reactions. Incubate for 1.5 h at 37 °C with shaking at approximately $150 \times g$.

8. Plate 150 μL of each transformation culture onto LB/ampicillin/IPTG/X-Gal plates. The cell pellets are obtained by centrifugation at $8000 \times g$ for 1 min, resuspended in 150 μL SOC medium. Incubate the plates overnight (i.e., 16–24 h) at 37 °C.

9. Use Plasmid Miniprep to isolate the recombinant plasmid DNA as instructed:

   (a) Resuspend pelleted bacterial cells in 250 μL buffer P1 and transfer to a microcentrifuge tube.

   (b) Add 250 μL buffer P2 and gently invert the tube 4–6 times to mix.

   (c) Add 350 μL buffer N3 and invert the tube immediately but gently 4–6 times.

   (d) Centrifuge for 10 min at $13,000 \times g$ in a table-top microcentrifuge.

   (e) Transfer supernatants to the QIAprep spin column by decanting.

   (f) Centrifuge for 60 s and discard the flow-through.

   (g) Wash the Column by adding 0.75 mL buffer PE and centrifuging for 60 s.

   (h) Discard the flow-through and centrifuge for an additional 1 min to remove residual wash buffer.

   (i) Place the column in a clean 1.5 mL microcentrifuge tube.

   (j) Add 50 μL water to the center of QIAprep and spin column to elute DNA.

   (k) Let stand for 1 min and centrifuge for 1 min.

10. For sequencing, use the purified recombinant plasmid DNA as templates and perform a cycle sequencing reaction using the BigDye Terminator V1.1 kit. For each reaction, mix 5 μL DNA template, 1.5 μL reaction mix, 3 μL 5× reaction buffer, 1.5 μL 10 μM T7 primer, and 4 μL sterile water. Mix briefly by pipetting and amplify DNA using the following PCR conditions: 1 cycle of denaturing at 1 min and 96 °C, 30 cycles of 10 s at 96 °C, 5 s, 50 s, 4 min at 55 °C and 1 cycle of 4 °C for 10 min.

11. After cycle sequencing, clean up the reaction by centrifuging the reaction plate and pipetting the SAM™ solution into each well (i.e., 20 μL per well for a 96-well plate).

12. Aspirate XTerminator solution and add to each well using a wide-bore pipette tip (i.e., 20 μL per well for a 96-well plate).

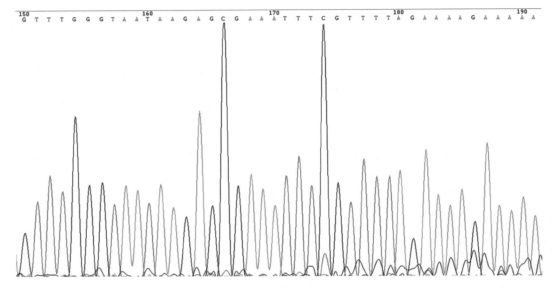

Fig. 4 Screenshots of bisulfite polymerase chain reaction sequencing of Cx43 gene promoter CpG islands. *Blue curve* indicates the peak of the sulfonated methylCpG site as C. *Red curve* demonstrates the sulfonated non-methylated CpG site as T. Reprinted with permission from ref. 19

13. Seal the plate using a clear adhesive film. Mix for 30 min and centrifuge the reaction plate briefly.

14. Perform sequencing using the DNA analyzer. Fragment analysis is applied to obtain the ratio of amplicons derived from bisulfite-converted methylated and unmethylated connexin DNA (Fig. 4).

3.4.3 MassArray Analysis

This method is ideal for detection of gene methylation, for discrimination between methylated and non-methylated samples, and for quantifying the methylation levels of DNA. In essence, cleavage products are generated for the reverse transcription reactions for both U (T) and C in separate reactions. Each cleavage product encloses either a CpG site, called a CpG unit, or an aggregate of multiple CpG sites [18]. For both T and C reactions, the resulting cleavage products have the same length and differ only in their nucleotide composition. A distinct signal pair pattern results from the methylated and non-methylated template DNA and is analyzed by the matrix-assisted laser desorption/ionization time-of-flight mass spectrometry technique (MALDI-TOF) (Fig. 5).

1. Perform PCR amplification of DNA following bisulfite modification using MassArray primers. Each PCR is split into two cleavage (i.e., T and C) reactions. The volume for single reaction is as follows: 0.5 μL 1× Hot Star buffer, 0.04 μL 200 μM dNTP mix, 0.04 μL 0.2 U/μL Hot Star Taq, 1 μL

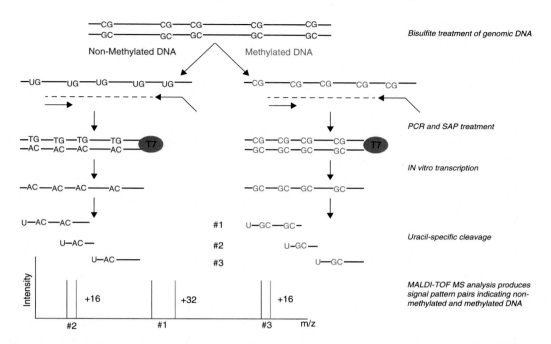

Fig. 5 Schematic diagram of the MassArray system. The bisulfite treatment converts non-methylated cytosine into uracil, thus generating methylation-dependent sequence changes in the genomic DNA template. PCR, with T7-promoter tagged reverse primers, is used to amplify the template, while preserving the induced sequence changes. These RNA products are processed by T-base-specific cleavage, yielding small RNA fragments

200 nM forward primer, 1 μL 200 nM reverse primer, and 1 μL DNA template.

2. Seal the plates and cycle as follows: 94 °C for 15 min, 45 cycles at 94 °C for 20 s, melting temperature of primer for 30 s, 72 °C for 1 min and 72 °C for 3 min.

3. Add 2 μL of shrimp alkaline phosphatase to each 5 μL PCR reaction to dephosphorylate unincorporated dNTPs from the PCR. Incubate the plates for 20 min at 37 °C and then incubate at 85 °C for 5 min.

4. Prepare transcription/RNase A cocktail for each cleavage reaction (i.e., T and C). The 5 μL total condition per plate is composed of 3.15 μL RNase-free double distilled water. 0.89 μL 0.64× T7 Polymerase buffer, 0.24 μL T/C cleavage mix, 0.22 μL 3.14 mM DTT, 0.44 μL 22 U T7 RNA and DNA Polymerase, and 0.06 μL 0.09 mg/mL RNase A.

5. Add 5 μL transcription/RNase A cocktail and 2 μL of shrimp alkaline phosphatase PCR sample into a new uncycled microtiter plate. Centrifuge the plates for 1 min and incubate the plates at 37 °C for 3 h.

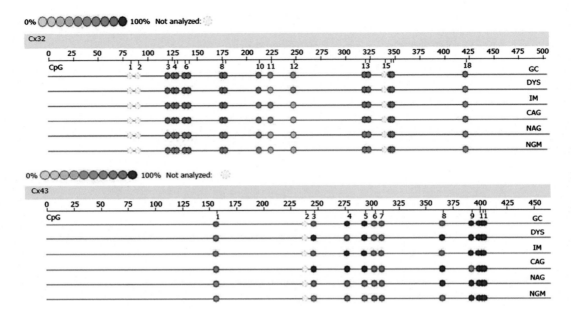

Fig. 6 Representative results for analysis of DNA methylation of Cx32 and Cx43 gene promoters at different gastric carcinogenesis stages by MassArray analysis. The validated length for Cx32 gene is 484 CpG sites containing 18 CpG sites (15 detected). The validated length for Cx43 gene is 447 bp with a total of 12 CpG sites (11 detected). This panel provides graphical representations of the CpG sites within the selected amplicon. The color code refers to the degree of DNA methylation shown in the methylation panel in order to provide quick and reliable comparison between samples and CpG sites. Reprinted with permission from ref. 19

6. Add 6 mg of clean resin to each well. Rotate for 10 min and spin down for 5 min at $3200 \times g$.

7. Add EpiTYPER reaction product and acquire spectra from the two cleavage reactions. The molecular weight of each fragment is determined by MALDI-TOF and the EpiTYPER software generates a report that contains quantitative information for each analyzed fragment (Fig. 6).

4 Notes

1. Alternatively, thawed tissue is introduced in the chilled nuclei lysis solution and homogenized for 10 s using a small homogenizer.

2. Some supernatant may remain near the pellets in the original tube containing the protein pellet. Leave this residual liquid in the tube to avoid contaminating the DNA solution with the precipitated protein.

3. The DNA pellet is very loose at this time. Using either a sequencing pipette tip or a drawn pipette is recommended to avoid aspirating the pellet into the pipette.

4. DNA protect buffer should turn from green to blue after addition to DNA bisulfite mix indicating sufficient mixing and accurate pH.

5. Carrier RNA is not necessary when using more than 100 ng DNA.

6. If there are precipitates in buffer BD, avoid transferring them to the spin column. Moreover, it is important to close the lid of the column before incubation.

7. To increase the yield of DNA in the eluate, transfer the spin column to a new 1.5 mL microcentrifuge tube, add an additional 20 µL buffer EB to the center of the membrane, and centrifuge for 1 min at $16,000 \times g$. Combine both eluates.

8. The methylated primer set assumes the CpGs are fully methylated; thus the primer will have all four bases in the sequence. The unmethylated primer set anneals to genomic DNA that is not methylated in the same primer binding site, and therefore will have T instead of C in the primer sequence. Design guidelines generally have a C or T near the 3' end of the two primer sets, respectively, where any mismatches will be discriminated against by the polymerase.

9. Design primers to the bisulfite-converted DNA sequence such that each primer is 25–30 bases long, has a melting temperature of approximately 60 °C, and does not hybridize to any CpG-cytosines.

10. It is important to test the primer sets with a control genomic DNA with a known methylation status along with genomic DNA with an unknown methylation status. The properly designed methylated primer set will only amplify the control methylated genomic DNA and not unmethylated genomic DNA, and the unmethylated primer set will only be positive for unmethylated genomic DNA.

11. It is recommended that DNA from peripheral blood is used as a control for the unmethylated reaction.

12. Ensure that temperatures 1–5 °C below and 1–5 °C above the Ta are tested in a gradient thermal cycler in order to determine the temperature at which a single specific PCR product is amplified. Using more cycles may be necessary if starting with less than 1 µg of genomic DNA.

Acknowledgements

The authors are grateful to the Endoscopic Unit of Third Xiangya Hospital of Central South University for the supply of clinical samples. This work was financially supported by the National Natural Science Foundation of China (No. 81172301).

References

1. Decrock E, Vinken M, De Vuyst E et al (2009) Connexin-related signaling in cell death: to live or let die? Cell Death Differ 16:524–536

2. Balda MS, Matter K (2000) The tight junction protein ZO-1 and an interacting transcription factor regulate ErbB-2 expression. EMBO J 19:2024–2033

3. Penes MC, Li X, Nagy JI (2005) Expression of zonula occludens-1 (ZO-1) and the transcription factor ZO-1-associated nucleic acid-binding protein (ZONAB)-MsY3 in glial cells and colocalization at oligodendrocyte and astrocyte gap junctions in mouse brain. Eur J Neurosci 22:404–418

4. Herrero-Gonzalez S, Gangoso E, Giaume C et al (2010) Connexin43 inhibits the oncogenic activity of c-Src in C6 glioma cells. Oncogene 29:5712–5723

5. Sirnes S, Lind GE, Bruun J et al (2015) Connexins in colorectal cancer pathogenesis. Int J Cancer 137:1–11

6. Dang X, Doble BW, Kardami E (2003) The carboxy-tail of connexin-43 localizes to the nucleus and inhibits cell growth. Mol Cell Biochem 242:35–38

7. Mesnil M, Krutovskikh V, Piccoli C et al (1995) Negative growth control of HeLa cells by connexin genes: connexin species specificity. Cancer Res 55:629–639

8. Leithe E, Rivedal E (2007) Ubiquitination of gap junction proteins. J Membr Biol 217:43–51

9. Vinken M, De Rop E, Decrock E et al (2009) Epigenetic regulation of gap junctional intercellular communication: more than a way to keep cells quiet? Biochim Biophys Acta 1795:53–61

10. Costello JF, Plass C (2001) Methylation matters. J Med Genet 38:285–303

11. Piechocki MP, Burk RD, Ruch RJ (1999) Regulation of connexin32 and connexin43 gene expression by DNA methylation in rat liver cells. Carcinogenesis 20:401–406

12. Yano T, Ito F, Kobayashi K et al (2004) Hypermethylation of the CpG island of connexin 32, a candidate tumor suppressor gene in renal cell carcinomas from hemodialysis patients. Cancer Lett 208:137–142

13. Yi ZC, Wang H, Zhang GY et al (2007) Downregulation of connexin 43 in nasopharyngeal carcinoma cells is related to promoter methylation. Oral Oncol 43:898–904

14. Tsujiuchi T, Shimizu K, Itsuzaki Y et al (2007) CpG site hypermethylation of E-cadherin and Connexin26 genes in hepatocellular carcinomas induced by a choline-deficient L-Amino Acid-defined diet in rats. Mol Carcinog 46:269–274

15. Hayatsu H, Wataya Y, Kai K et al (1970) Reaction of sodium bisulfite with uracil, cytosine, and their derivatives. Biochemistry 9:2858–2865

16. Kai K, Tsuruo T, Hayatsu H (1974) The effect of bisulfite modification on the template activity of DNA for DNA polymerase I. Nucleic Acids Res 1:889–899

17. Pappas JJ, Toulouse A, Bradley WE (2009) A modified protocol for bisulfite genomic sequencing of difficult samples. Biol Proced Online 11:99–112

18. Jurinke C, Denissenko MF, Oeth P et al (2005) A single nucleotide polymorphism based approach for the identification and characterization of gene expression modulation using MassARRAY. Mutat Res 573:83–95

19. Wang Y, Huang LH, Xu CX et al (2014) Connexin 32 and 43 promoter methylation in Helicobacter pylori-associated gastric tumorigenesis. World J Gastroenterol 20:11770–11779

Chapter 3

Detection of Connexins in Liver Cells Using Sodium Dodecyl Sulfate Polyacrylamide Gel Electrophoresis and Immunoblot Analysis

Joost Willebrords, Michaël Maes, Sara Crespo Yanguas, Bruno Cogliati, and Mathieu Vinken

Abstract

Since connexin expression is partly regulated at the protein level, immunoblot analysis represents a frequently addressed technique in the connexin research field. The present chapter describes the setup of an immunoblot procedure, including protein extraction and quantification from biological samples, gel electrophoresis, protein transfer, and immunoblotting, which is optimized for analysis of connexins in liver tissue. In essence, proteins are separated on a polyacrylamide gel using sodium dodecyl sulfate followed by transfer of proteins on a nitrocellulose membrane. The latter allows specific detection of connexins with antibodies combined with revelation through enhanced chemiluminescence.

Key words Connexins, Protein extraction, Sodium dodecyl sulfate polyacrylamide gel electrophoresis, Immunoblot analysis, Antibody, Enhanced chemiluminescence

1 Introduction

Renart, Reiser, and Stark were the firsts to describe a rapid and sensitive method for the separation and detection of proteins using diffusion over polyacrylamide gels and antibodies [1]. Later on, Towbin reported transfer of proteins in an electrical field [2]. This approach was optimized by Brunette, who introduced the designation western blot, playing on the Southern blot and northern blot methods used to detect DNA and RNA, respectively [3]. Throughout the years, the immunoblot technique underwent significant advances with respect to the sensitivity, robustness, and flexibility [4]. At present, the western blot technique is widely used because of its specific advantages in comparison with other protein detection methods, such as mass spectrometry and enzyme-linked immunosorbent assay

Mathieu Vinken and Scott R. Johnstone (eds.), *Gap Junction Protocols*, Methods in Molecular Biology, vol. 1437,
DOI 10.1007/978-1-4939-3664-9_3, © Springer Science+Business Media New York 2016

(ELISA). Overall, the immunoblot procedure has a great specificity and is cheap in comparison with mass spectrometry. Moreover, western blotting can detect proteins in small amounts of sample. On the other hand, semi-quantitative densitometric analysis is typically used to quantify protein expression on a blot, being less accurate than the ELISA detection method [5]. Connexin (Cx) half-life is very short (i.e., 1.5–5 h) compared to other transmembrane proteins [6–10]. Connexin mRNA translation can be modified through microRNAs or RNA-binding proteins and may be altered through trafficking and degradation mechanisms [11, 12].

As such, six connexin proteins gather to form a hemichannel in a single cell and two hemichannels of adjacent cells can connect to establish gap junctions [13]. The latter are important communication channels between neighboring cells, providing the passage of a number of molecules, such as cyclic adenosine monophosphate, adenosine triphosphate, inositol triphosphate, and ions, all of which play important roles in liver homeostasis [14–16]. The liver was the first organ in which connexins have been identified and hence several research efforts over the last decades have been focused on the analysis of these particular gap junction proteins [17–20]. In this regard, detection of connexin proteins using immunoblot analysis is a valuable method in liver research.

In this chapter, an immunoblot procedure optimized for analysis of hepatic connexins, specifically Cx26, Cx32, and Cx43, is outlined. The procedure basically consists of two steps, namely sodium dodecyl sulfate (SDS) polyacrylamide gel electrophoresis (PAGE) and immunoblotting. Adding SDS, a strong anionic detergent, to protein extracts will result in an equal distribution in charge per unit mass. When applying voltage to the polyacrylamide gel, proteins will migrate at different speed and will be separated according to their molecular weight [21]. Subsequently, separated proteins can be transferred onto a membrane, typically made of nitrocellulose, polyvinylidene difluoride, or polyvinylpyrrolidone, upon establishment of an electrical field [22]. In order to reduce nonspecific binding of antibodies on the membrane in the next step, blots are incubated in a solution containing bovine serum albumin (BSA), nonfat dry milk, or casein [23]. Visualization of the protein can be achieved by a number of methods, of which enhanced chemiluminescence (ECL) is the most popular one. ECL is a very sensitive method and can be used for relative quantification of protein abundance on the membrane by detecting horseradish peroxidase from antibodies. The reaction product produces luminescence (Fig. 1), which can be captured on photographic film, and this is correlated with the amount of target protein on the membrane [24].

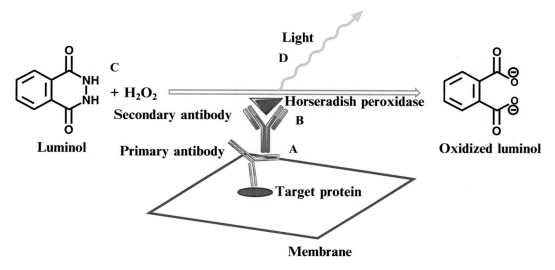

Fig. 1 Principle of ECL detection method. (**a**) Primary antibody binds to target protein on the membrane. (**b**) Secondary antibody, linked to horseradish peroxidase, binds to primary antibody. (**c**) Horseradish peroxidase and H_2O_2 aid in the oxidation of luminol. (**d**) This reaction emits light, which correlates to the amount of protein on the membrane

2 Materials

All solutions should be prepared with ultrapure water (i.e., purified and deionized water with a constant sensitivity of 18 MΩ at 25 °C).

2.1 Protein Extraction

1. Ethylenediaminetetraacetic acid (EDTA) solution: 0.5 M EDTA in water. Adjust to pH 8.0 (*see* **Note 1**). This solution can be stored for 6 months at 4 °C.

2. Lysis buffer (*see* **Note 2**): *Prior* to use, add 100 μL EDTA solution and 100 μL 100× protease and phosphatase inhibitor cocktail to 9.8 mL radio-immunoprecipitation assay buffer.

3. Centrifuge suitable for cooling at 4 °C.

4. Sonicator (i.e., for cell culture samples) or mixer (i.e., for liver tissue samples).

5. Rotator.

2.2 Protein Quantification Assay (See Note 3)

1. Bicinchoninic acid (BCA) working reagent: *Prior* to use, mix BCA reagent A with BCA reagent B in a ratio 50:1. The solution turns blue. BCA working reagent can be stored for several days in a closed container at room temperature.

2. BSA standards (Table 1). Keep on ice.

3. Flat-bottom 96-well plate.

4. Multiplate reader.

5. Incubator.

Table 1
BSA standard dilutions

BSA standard	Water (µL)	Stock (µL)	Final concentration (µg/mL)
A	0	300 from stock	2000
B	125	375 from stock	1500
C	325	325 from stock	1000
D	175	175 from B	750
E	325	325 from C	500
F	325	325 from E	250
G	325	325 from F	125
H	400	100 from G	25

BSA bovine serum albumin

2.3 Preparation of Gels (Table 2)

1. Tris buffer 1: 1.5 M Tris in water. Adjust to pH 8.8. This solution can be stored for 6 months at 4 °C.

2. Tris buffer 2: 0.5 M Tris in water. Adjust to pH 6.8. This solution can be stored for 6 months at 4 °C.

3. Polyacrylamide solution: 30 % acrylamide/bisacrylamide 37.5:1 solution (*see* **Note 4**).

4. SDS solution: 10 % SDS in water. This solution can be stored for 6 months at room temperature.

5. Ammonium persulfate (APS) solution: 10 % APS in water. The solution is prepared *ex tempore* and should be kept on ice.

6. Tetramethylethylenediamine (TEMED).

7. Isobutanol solution: 1:3 isobutanol/water solution. Shake and let the solution rest after preparation for at least 30 min.

8. Glass plates.

9. Combs.

10. Filter paper.

2.4 SDS-PAGE

1. Bromophenol blue solution: 0.5 % bromophenol blue in water. This solution can be stored for 6 months at 4 °C.

2. Sample buffer: Mix 2.5 mL Tris buffer 2, 2 mL SDS solution, 2 mL glycerol, 200 µL bromophenol blue solution, and 2.35 mL water. This solution can be stored at −20 °C for 2 weeks. 1/10 β-mercaptoethanol is added *prior* to use (*see* **Note 5**).

3. Running buffer: 24.9 mM Tris, 192 mM glycine, 3.5 mM SDS in water. About 1 L per 2 gels is needed. Prepare *ex tempore*.

4. Protein marker (*see* **Note 6**).

5. Electrophoresis tank.

6. Power supply source.

2.5 Protein Transfer

1. Transfer buffer: 24.9 mM Tris, 192 mM glycerol, 20 % methanol in water. Prepare *ex tempore*.

2. Nitrocellulose membrane (*see* **Note 7**).

3. Ponceau S solution: 0.03 M Ponceau S, 1.84 M trichloroacetic acid, and 1.18 M 5-sulfosalicylic acid dihydrate in water. This solution can be reused when stored for 6 months at room temperature.

4. Electrophoresis tank.

5. Power supply source.

2.6 Immunoblotting and Protein Detection

1. 10× Tween-supplemented Tris-buffered phosphate-buffered saline (TBST): 0.20 M Tris, 1.36 M NaCl, 9 mM Tween 20 in water. Adjust to pH 7.6. This solution can be stored for 6 months at 4 °C.

2. 1× TBST: Dilute 10× TBST ten times in water. Prepare *ex tempore*.

3. Blocking buffer: 5 % milk powder in 1× TBST (*see* **Note 8**). Prepare *ex tempore*.

4. Appropriate primary and secondary antibody: Polyclonal anti-Cx26 antibody produced in rabbit (Thermo Scientific, USA), polyclonal anti-Cx32 primary antibody produced in rabbit (Sigma, USA), polyclonal anti-Cx43 primary antibody produced in rabbit (Sigma, USA). Polyclonal goat anti-rabbit immuno-globulins/horseradish peroxidase secondary antibody (Dako, Denmark) (*see* **Note 9**).

5. Pierce™ ECL western blotting substrate (Thermo Scientific, USA): Mix 2 mL peroxide solution with 2 mL luminol enhancer solution *ex tempore*.

6. Fixation solution.

7. Development solution.

8. Photographic films.

2.7 Stripping of the Immunoblot

Restore™ western blot (Thermo Scientific, USA).

3 Methods

3.1 Protein Extraction

1. Add 10 µL of lysis buffer per mg liver tissue or 100 µL per cell culture dish with a diameter of 10 cm (*see* **Note 10**).

2. For cultured cells, sonicate 30 s with 50 % pulse while keeping the cells on ice (*see* **Note 11**). For liver tissue samples, homogenize using a mixer and keep on ice.

Table 2
Composition of stacking and separating polyacrylamide gels

	Separation gel (7.5%)	Separation gel (10%)	Separation gel (12%)	Separation gel (15%)	Stacking gel (4%)
Water (mL)	4.85	4.00	3.35	2.35	3.05
1.5 M Tris buffer (mL)	2.50	2.50	2.50	2.50	–
0.5 M Tris buffer (mL)	–	–	–	–	1.25
Polyacrylamide solution (mL)	2.50	3.33	4.00	5.00	0.65
SDS solution (mL)	0.10	0.10	0.10	0.10	0.050
APS solution (mL)	0.050	0.050	0.050	0.050	0.025
TEMED (mL)	0.015	0.015	0.015	0.015	0.015

SDS Sodium dodecyl sulfate, *APS* ammonium persulfate, *TEMED* tetramethylethylenediamine

3. Shake for 15 min on a rotator at 4 °C (*see* **Note 12**).

4. Centrifuge at $14{,}000 \times g$ for 15 min at 4 °C.

5. Transfer the supernatant to a new tube and store at –80 °C.

3.2 Protein Quantification Assay

1. All samples, standards, and blanks should be measured in triplicate and should be kept on ice.

2. Prepare the BSA standards (Table 1) and the BCA working reagent.

3. Pipet 10 μL of each properly diluted sample (*see* **Note 13**) and BSA standards in the respective wells of a flat-bottom 96-well plate. Add 200 μL of the BCA working reagent to each well. Use 210 μL BCA working reagent as a blank. Mix the plate thoroughly on a plate shaker for 30 s.

4. Protect the plate from light and incubate at 37 °C for 30 min.

5. Cool the plate to room temperature and measure the absorbance at 562 nm on the multiplate reader.

6. Subtract the average absorbance measurement of the blank standard from the measurements of all other standards and samples. Prepare a standard curve by plotting the average absorbance measurement for each BSA standard as a function of concentration expressed in μg/mL. Use the standard curve to determine the protein concentration of each sample by interpolation.

3.3 Preparation of Gels (See Note 14)

1. Prepare the gel solution (*see* **Note 15**) in a conical flask (Table 2).

2. Pour the separation gel solution between the glass plates and add a thin layer of isobutanol solution. The polymerization process starts instantly. Let rest for 30 min at room temperature.

3. Prepare the stacking gel solution (Table 2).

4. Remove the isobutanol solution using a filter paper, wash three times with water, and pour the stacking gel solution on top of the separation gel. Correctly place the combs into the gel and let rest for at least 30 min at room temperature.

5. Remove the combs and rinse the wells three times with water.

3.4 SDS-PAGE

1. Thaw the protein samples on ice before starting.

2. Calculate the volume needed to load 50 μg protein from each sample (*see* **Note 16**) on the gel based on the respective protein concentration. Dilute until 10 μL with lysis buffer and add 10 μL sample buffer. All samples should be kept on ice during preparation.

3. Install the gel cassette into the electrophoresis tank (*see* **Note 17**) and add running buffer between the gels and in the tank (*see* **Note 18**). Spin the samples shortly before loading on the gel and shortly mix protein ladder solution. Load 5 μL of the protein ladder (*see* **Note 6**) in the first well and load samples (i.e., 20 μL) in the other wells. Empty wells are loaded with a mixture of 10 μL lysis buffer and 10 μL sample buffer (*see* **Note 19**).

4. Close the electrophoresis tank, connect to the voltage source and start separation by applying a constant voltage of 150 V. Stop when the bromophenol blue line reaches the end of the gel (*see* **Note 20**).

3.5 Protein Transfer (See Note 21)

1. Cut a nitrocellulose membrane fit to the size of the gel and let soak in cold transfer buffer together with the blotting paper and sponges.

2. Take the cassette out of the tank, open and remove the stacking gel using a spatula (*see* **Note 22**). Put the separation gel in the transfer recipient with the nitrocellulose membrane underneath. Add blotting paper on the membrane and on the gel. Keep in mind that the gel is the top layer while the membrane forms the bottom layer. Remove all air bubbles by rolling over using a roller. Add a sponge to each side and put in a cassette. Add transfer buffer to the tank and apply a constant voltage of 150 V for 1 h (*see* **Note 23**).

3. Remove the membrane from the cassette and incubate with Ponceau S solution for 3 min while shaking. Ponceau S will bind to all proteins and allows assessing the efficacy of the transfer and checking equal loading (*see* **Note 24**).

4. Rinse the membrane three times with water and shake with TBST until all Ponceau S is removed from the membrane.

3.6 Immunoblotting and Protein Detection

1. Block the nitrocellulose membrane for 1 h in blocking buffer at room temperature on a shaker. Make sure that the blotting side is faced up.

2. Incubate with the primary antibody (*see* **Note 25**) diluted in blocking buffer overnight at 4 °C on a shaker (*see* **Note 26**).

3. Wash the membrane three times 10 min with TBST on a shaker.

4. Incubate with the secondary antibody solution for 1 h at room temperature (*see* **Note 27**).

5. Wash three times 10 min with TBST on a shaker (*see* **Note 28**).

6. Prepare the Pierce™ ECL western blotting substrate solution and incubate with the membrane for 1 min at room temperature.

7. Place the photographic film on the membrane and close the cassette for an appropriate period of time (*see* **Note 29**).

8. Remove the film and place for 1 min in fixation solution, rinse shortly with water, and expose for 1 min to development solution. Rinse shortly with water and let dry (*see* **Note 30**).

3.7 Stripping of the Immunoblot (See Note 31)

1. If necessary, rehydrate the membrane by immersing in TBST.

2. Incubate the membrane in Restore™ western blot stripping buffer for 5–15 min at room temperature.

3. Wash three times 5 min with TBST on a shaker.

4. Check the efficiency of the stripping process by re-incubation of the membrane with the previously used secondary antibody, followed by washing with TBST and developing of the immunoblot.

5. If no signal is observed, proceed with re-incubation of the membrane with the antibody of interest.

3.8 Processing of Results

1. If stripping of the membrane is necessary (*see* **Note 31**), redevelop the membrane as described in Subheading 3.6 with appropriate primary and secondary antibody of a suitable housekeeping protein (*see* **Note 32**).

2. Analyze and quantify the protein signals on the membrane using Quantity One 1D Analysis Software (Bio-Rad, USA) by semi-quantitative densitometry for both the target proteins and housekeeping proteins (*see* **Note 33**).

3. Correct the target protein measurements to the housekeeping protein results in each lane of the membrane (*see* **Note 33**).

4. Statistically process the results using an appropriate test (*see* **Note 34**).

4   Notes

1. pH adjustment of a solution is commonly performed by using HCl. A concentrated HCl solution (i.e., 37 %) can be used for this purpose, yet it can be useful to use a less concentrated HCl solution (i.e., 5 %) to avoid a sudden drop in pH.

2. The selection of an appropriate lysis buffer is of utmost importance. It should be optimally designed to efficiently extract the protein(s) of interest, whilst leaving their antigenic sites intact. Particularly the use of a suitable detergent for protein extraction and solubilization should be taken into account as well as addition of a proper set of protease and phosphatase inhibitors [25]. Some antibodies will only recognize a protein in its native non-denatured form and will not detect a protein that has been extracted with a denaturing detergent, such as SDS, deoxycholate, Triton X-100, and NP-40. In general, the following range of variables should be optimized: salt concentration 0–1 M, non-ionic detergent 0.1–2 %, ionic detergents 0.01–0.5 %, divalent cation concentration 0–10 mM, EDTA concentration 0–5 mM and pH 6–9 [26].

3. A diverse array of protein quantification assays is currently available. It is of major importance to select the correct assay based on the type of sample and concentration range of protein in the sample, since each test has its specific limitations [27, 28]. For cultured hepatocytes and freshly isolated liver tissue, it is recommended to use the BCA assay, as it provides a good combination of specificity and ease of use. The BCA assay is a spectrophotometric assay based on the alkaline reduction of cupric ions to cuprous ions mediated by proteins. This is followed by chelation and color development by the BCA reagent. BSA is used as a standard because of its great working range and stability [29]. Immunoglobulin G could also be used as a standard against which the concentration of protein in samples is normalized [30].

4. Use gloves and be careful when handling acrylamide, as it is a neurotoxic and carcinogenic agent [31–33].

5. Add β-mercaptoethanol to the sample buffer in a well-ventilated area, as it is considered toxic and can cause irritation to the nasal cavity, respiratory tract, and skin.

6. A molecular weight marker should be loaded onto the gel along with the analyzed samples in order to enable the determination of the protein size and to monitor the progress of the electrophoretic run. Precision Plus Protein™ Kaleidoscope™ Prestained Standards (Bio-Rad, USA) is a suitable protein marker for the detection of connexin proteins, since it allows monitoring of

electrophoretic separation, molecular weight seizing, and blot transfer efficacy on SDS-PAGE gels by fluorescence. The choice of an appropriate molecular weight marker depends on the protein detection method and molecular weight range of the connexin of interest. Bio-Rad (USA) provides a series of protein markers, such as prestained protein ladders for visual assessment during electrophoresis and transfer, unstained protein ladders for coomassie blue and zinc staining, and western blotting standards for fluorescent visualization and colorimetric or chemiluminescent immunodetection on western blots.

7. There are three types of blotting membranes available, namely nitrocellulose, which is the most popular one, polyvinylidene fluoride, which requires pretreatment in methanol for 1–2 min, and polyvinylpyrrolidone. Nitrocellulose membranes are typically used for low molecular weight protein detection, while polyvinylpyrrolidone is more suitable for northern blotting. All types of membranes can be stripped and reprobed.

8. When preparing blocking buffer, make sure that all milk powder is dissolved by stirring the solution for about 20 min. This solution can be used up to 5 days when stored at 4 °C. It is recommended to mix again before use.

9. The decision regarding whether to use polyclonal or monoclonal antibodies depends on a number of factors. Polyclonal antibodies can be generated more rapidly and at less expense, and are more stable over a broad pH and salt concentration range. In contrast, monoclonal antibodies are homogenetic, which can be useful for detection of changes in phosphorylation status. However, small changes in the structure of the epitope can affect the effectiveness of monoclonal antibodies [34].

10. For measurement of connexins via immunoblotting, it is strongly recommended to use 10 cm diameter cell culture dishes with hepatocytes plated at a density of 0.57×10^5 cells/ cm^2 (i.e., a total of 4.4×10^6 hepatocytes) in order to obtain sufficient amounts of total cellular protein. 70–100 µL of lysis buffer is recommended for a total amount of 4.4×10^6 hepatocytes. It is advised to start with a lower volume of the lysis buffer and to observe the color of cell homogenate upon resuspending the cell pellet. Extremes, such as transparency or brownish color of the cell lysate, indicating too high or too low dilution respectively, should be avoided. For liver tissue, 10 µL lysis buffer per mg should be sufficient.

11. Keeping the cells on ice will prevent protein degradation by heat produced from sonication.

12. Shaking on a rotator is not strictly necessary, but it increases the yield of the protein extraction.

13. Dilution of protein extracts for quantification with the BCA assay needs to be optimized. When adding 10 µL lysis buffer per mg tissue or when using 100 µL lysis buffer for a total amount of 4.4×10^6 hepatocytes cultured in 10 cm diameter cell culture dishes, a 1/20 or 1/40 dilution is usually sufficient.

14. Appropriate results have been obtained with PROTEAN® TGX Stain-Free™ Precast gels (Bio-Rad, USA) for the detection of connexin proteins. These gels are based on a novel modification of the Laemmli system (i.e., Tris, glycerol, and SDS), which increases gel matrix stability, allows the use of Tris-glycerol-SDS running buffer, and offers rapid UV detection of connexin proteins using a stain-free imaging system, such as ChemiDoc™ MP. Moreover, an optimal resolution of connexins is obtained in the 10–100 kDa molecular weight range. The Laemmli system operates at pH 9.5, which can cause deamination and alkylation of proteins. Therefore, the Bis-Tris gel system can be a suitable alternative, as this system has a lower pH of 7.0. The Bis-Tris system ensures increased protein stability, decreased protein modifications, and a faster running time [35].

15. Depending on the connexin of interest, a different polyacrylamide percentage is considered during gel preparation, in particular 10 % gel for Cx43, 12 % gel for Cx32 and Cx26 (Table 2). It is recommended to add TEMED and APS solution as final ingredients, as they initiate gel formation instantly.

16. The total amount of protein loaded onto the gel must be optimized. It is dictated by cellular abundance of the protein of interest as well as reactivity of the primary antibody used. Overall, the scarcer the protein and/or the lower affinity of the antisera, the higher the amount of total cellular proteins that should be loaded in order to avoid false negatives. It is recommended to start with 50 µg total cellular proteins per lane and to include a positive (*e.g.*, transfected cell line) and a negative (*e.g.*, tissue from connexin-deficient animals) control. Exceeding this amount should be avoided, as this will result in gel overloading and creation of protein "smudges" instead of bands.

17. When using PROTEAN® TGX Stain-Free™ Precast Gels (Bio-Rad, USA), gently remove the combs and the green tape from the gel cassette before use.

18. Make sure that the running buffer between the cassettes and in the electrophoresis tank does not make contact with each other, as this may interfere with the voltage potential.

19. It can be useful to add a negative control between the weight marker and the first sample, as there might be some interference. It is critical to fill all the wells of the gel. If not, this could lead to so-called smiling of the pattern.

20. The running time should be optimized and will depend on the molecular weight of the connexin under investigation. The goal is to spread as much as possible the molecular weight area where the protein of interest is located.

21. The traditional wet blotting method requires water, which helps to prevent the generation of oxygen from the copper anode resulting in a more consistent protein transfer without causing band distortion. The alternative semi-dry blotting uses high-field strength and currents to reduce the transfer time of proteins from gels onto membranes within 7 min [28].

22. When using PROTEAN® TGX Stain-Free™ Precast Gels (Bio-Rad, USA), carefully open the cassette using a spatula and check the loading of the gel by UV detection with ChemiDoc™ MP using the stain-free activation. The latter allows total protein staining which can serve as superior alternative for loading control with housekeeping proteins [36].

23. Transfer of connexin proteins can be efficiently performed by using the Trans-Blot® Turbo™ Transfer System (Bio-Rad, USA), which allows rapid and efficient transfer, and the Trans-Blot® Turbo™ Mini Nitrocellulose Transfer Pack (Bio-Rad, USA), which contains an appropriate buffer, membrane, and filter paper. High molecular weight proteins (i.e., 200–250 kDa) can sometimes remain on the gel because of the rapid transfer. This usually poses no problem for the detection of connexin proteins, as their molecular weight ranges from 26 to 60 kDa. Transfer time of connexins depends on the molecular weight of the protein to be analyzed. As a rule of thumb, the higher the molecular weight, the higher the voltage and the longer the running time. It is also critical to reduce the production of heat during immunoblotting, as this may harm the integrity of the proteins. This can be done by adding a cooling element.

24. Not only transfer efficacy can be tested with Ponceau S. Indeed, it also has shown to be a good loading control even superior to housekeeping proteins [36].

25. A plethora of companies have primary connexin antibodies available, including Sigma, Abcam, Thermo Scientific, Santa Cruz, EMD Millipore, Genetex, ProteinTech, Alomone, Biocompare, Life Technologies, and Merck Millipore.

26. Concentrations of both primary and secondary antibody solutions should be optimized for each connexin of interest. It is recommended to test different concentrations of the primary antibody and keep the concentration of the secondary antibody low to avoid nonspecific binding. Other possibilities to be included during optimization are incubation for 1 h at 37 °C or 2 h at room temperature.

27. Most commonly used are secondary antibodies linked to horse-radish peroxidase, an enzyme which catalyzes the oxidation of luminol, producing light [37]. However, other detection methods are also available, such as the use of a fluorochrome or a colorimetric substrate [38]. Detection via fluorescence is performed if high sensitivity is required, typically when there are two different targets on one blot. Dyes like fluorescein and rhodamine are often used, but have numerous limitations, including relatively low fluorescence intensity and tendency to photobleach. Colorimetry is simple and the most cost-effective method of detection. Commonly used substrates include 5-bromo-4-chloro-3-indolylphosphate-p-toluidine salt and nitro-blue tetrazolium chloride [39].

28. Inefficient washing was shown to be the primary reason for high background in western blots. Lowering the non-ionic detergent (i.e., Tween 20) concentration in antisera dilution buffer (i.e., 0.02–0.005 %) and rising the concentration of the detergent in the wash buffer (i.e., 0.5 %) could improve the results [40].

29. The exposure time of the membrane to the film must be experimentally optimized. It is important that several exposures are produced to ensure that the response is within a linear range of the photographic film. One should aim at obtaining the sharpest bands possible that still have a greyish appearance. This is one of the advantages of the ChemiDoc™ MP (Bio-Rad, USA), which allows detection of overexposed bands.

30. Cx43 is the only connexin protein for which posttranslational phosphorylation can be detected in western blotting using anti-Cx43 primary antibody produced in rabbit (Sigma, USA) (Fig. 2). Additionally, some connexins undergo partial proteolysis during protein extraction, which may lead to detection at other molecular weight values than suggested by the designation. The best example is Cx32, for which cDNA sequencing predicts a molecular weight of 32 kDa, but that is usually found at 27 kDa in immunoblot analysis (Fig. 2).

31. Stripping of the membrane should be avoided as it results in gradual loss of blotted proteins from the membrane. It is therefore advised to directly proceed with re-incubation of the membrane, unless the protein of interest and the loading control protein have approximately the same molecular weight or when species-based interference is expected between antisera used for the detection of the target protein and the loading control. When using the stain-free loading control, stripping of the membrane is not necessary since no housekeeping genes are used.

32. It is advisable to proceed with re-incubation of the membrane with an antibody directed against a typical housekeeping protein, such as glyceraldehyde 3-phosphate dehydrogenase (GAPDH),

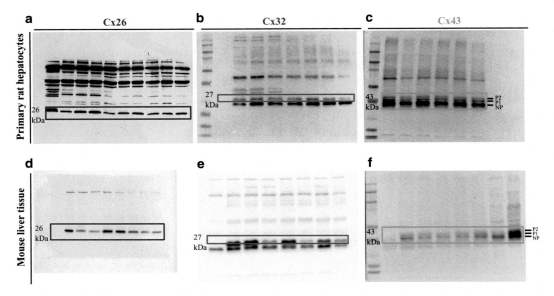

Fig. 2 Detection of Cx26, Cx32, and Cx43 in cultured primary rat hepatocytes and mouse liver tissue by means of western blotting. (**a**) Cx26 detected in primary rat hepatocytes using 12 % polyacrylamide gel, 1/250 anti-Cx26 polyclonal antibody produced in rabbit (Thermo Scientific, USA), 1/2500 polyclonal goat anti-rabbit immunoglobulins/horseradish peroxidase (Dako, Denmark), nitrocellulose membrane, and Pierce™ ECL western blotting substrate (Thermo Scientific, USA). (**b**) Cx32 detected in primary rat hepatocytes using Mini-PROTEAN® TGX 12 % Stain-Free™ Precast gels (Bio-Rad, USA), 1/1000 anti-Cx32 polyclonal antibody produced in rabbit (Sigma, USA), 1/1000 polyclonal goat anti-rabbit immunoglobulins/horseradish peroxidase (Dako, Denmark), nitrocellulose membrane, and Pierce™ ECL western blotting substrate (Thermo Scientific, USA). (**c**) Three forms of Cx43 detected in primary rat hepatocytes, using Mini-PROTEAN® TGX 10 % Stain-Free™ Precast gels (Bio-Rad, USA), 1/1000 anti-Cx43 polyclonal antibody produced in rabbit (Sigma, USA), 1/1000 polyclonal goat anti-rabbit immunoglobulins/horseradish peroxidase (Dako, Denmark), nitrocellulose membrane, and Pierce™ ECL western blotting substrate (Thermo Scientific, USA). (**d**) Cx26 detected in mouse liver tissue using 12 % polyacrylamide gel, 1/1000 anti-Cx26 polyclonal antibody produced in rabbit (Thermo Scientific, USA), 1/1000 polyclonal goat anti-rabbit immunoglobulins/horseradish peroxidase (Dako, Denmark), nitrocellulose membrane, and Pierce™ ECL western blotting substrate (Thermo Scientific, USA) [46]. (**e**) Cx32 detected in mouse liver tissue, using Mini-PROTEAN® TGX 12 % Stain-Free™ Precast gels (Bio-Rad, USA), 1/1000 anti-Cx32 polyclonal antibody produced in rabbit (Sigma, USA), 1/1000 polyclonal goat anti-rabbit immunoglobulins/horseradish peroxidase (Dako, Denmark), nitrocellulose membrane, and Pierce™ ECL western blotting substrate (Thermo Scientific, USA). (**f**) 3 forms of Cx43 detected in mouse liver tissue, using Mini-PROTEAN® TGX 10 % Stain-Free™ Precast gels (Bio-Rad, USA), 1/1000 anti-Cx43 polyclonal antibody produced in rabbit (Sigma, USA), 1/1000 polyclonal goat anti-rabbit immunoglobulins/horseradish peroxidase (Dako, Denmark), nitrocellulose membrane, and Pierce™ ECL western blotting substrate (Thermo Scientific, USA) (NP, non-phosphorylated Cx43 variant; P1, P1 phosphorylated Cx43 variant; P2, P2 phosphorylated Cx43 variant)

β-actin, or α-tubulin (Table 3), which serves as a loading control. Housekeeping proteins should show constant protein expression patterns between different tissue types. It is required to check if all lanes in the gel were evenly loaded with sample, especially when a comparison must be made between the expression levels of a protein in different samples. Detection of housekeeping pro-

Table 3
Most commonly used loading controls

Loading control	Sample type	Molecular weight (kDa)	Remarks
Beta actin	Whole cell/cytoplasmic	43	Not suitable for skeletal muscle samples. Changes in cell growth conditions and interactions with extracellular matrix components may alter actin protein synthesis [41]
GAPDH		37	Some physiological factors, such as hypoxia and diabetes, increase GAPDH expression in certain cell types [42, 43]
Tubulin		55	Tubulin expression may vary according to resistance to antimicrobial and antimitotic drugs [44, 45]
VDAC1/Porin	Mitochondrial	31	
COXIV		16	
Lamin B1	Nuclear	66	Not suitable for samples devoid of nuclear envelope
TBP		38	Not suitable for samples devoid of DNA

COXIV cytochrome c oxidase subunit IV, *GAPDH* glyceraldehyde 3-phosphate dehydrogenase, *TBP* TATA binding protein, *VDAC* voltage-dependent anion channel

teins is also useful to check for equal transfer from the gel to the membrane across the whole gel. The loading control bands can be used to quantify the protein amounts in each lane by means of densitometry.

33. From a statistical point of view, it is strongly recommended to include at least three biological and three technical repeats. With exposures captured on photographic film, one can utilize densitometric methods to measure relative quantities of a target protein as well as loading control protein and subsequently compare to a control (*e.g.*, specific time point or untreated sample). Bio-Rad Quantity One and Image Lab™ Software (Bio-Rad, USA) are available for image analysis of bands on photographic film and PROTEAN® TGX Stain-Free™ Precast Gels (Bio-Rad, USA) respectively.

34. Results can be processed and evaluated by 1-way analysis of variance followed by post hoc Bonferroni tests for testing of three or more groups. When comparing only two groups, a *t*-test is sufficient.

Acknowledgements

This work was financially supported by the University Hospital of the Vrije Universiteit Brussel-Belgium (Willy Gepts Fonds UZ-VUB), the Fund for Scientific Research-Flanders (FWO grants G009514N and G010214N), the European Research Council (ERC Starting Grant 335476), the University of São Paulo-Brazil, and the Foundation for Research Support of the State of São Paulo (FAPESP SPEC grant 2013/50420-6).

References

1. Renart J, Reiser J, Stark GR (1979) Transfer of proteins from gels to diazobenzyloxymethyl-paper and detection with antisera: a method for studying antibody specificity and antigen structure. Proc Natl Acad Sci U S A 76:3116–3120

2. Towbin H, Staehelin T, Gordon J (1979) Electrophoretic transfer of proteins from polyacrylamide gels to nitrocellulose sheets: procedure and some applications. Proc Natl Acad Sci U S A 76:4350–4354

3. Burnette WN (1981) "Western blotting": electrophoretic transfer of proteins from sodium dodecyl sulfate-polyacrylamide gels to unmodified nitrocellulose and radiographic detection with antibody and radioiodinated protein A. Anal Biochem 112:195–203

4. Dubitsky A, DeCollibus D, Ortolano GA (2002) Sensitive fluorescent detection of protein on nylon membranes. J Biochem Biophys Methods 51:47–56

5. McDonough AA, Veiras LC, Minas JN et al (2015) Considerations when quantitating protein abundance by immunoblot. Am J Physiol Cell Physiol 308:C426–C433

6. Beardslee MA, Laing JG, Beyer EC et al (1998) Rapid turnover of connexin43 in the adult rat heart. Circ Res 83:629–635

7. Musil LS, Le AC, VanSlyke JK et al (2000) Regulation of connexin degradation as a mechanism to increase gap junction assembly and function. J Biol Chem 275:25207–25215

8. Darrow BJ, Laing JG, Lampe PD et al (1995) Expression of multiple connexins in cultured neonatal rat ventricular myocytes. Circ Res 76:381–387

9. Fallon RF, Goodenough DA (1981) Five-hour half-life of mouse liver gap-junction protein. J Cell Biol 90:521–526

10. Laird DW, Puranam KL, Revel JP (1991) Turnover and phosphorylation dynamics of connexin43 gap junction protein in cultured cardiac myocytes. Biochem J 273:67–72

11. Su V, Lau AF (2014) Connexins: mechanisms regulating protein levels and intercellular communication. FEBS Lett 588:1212–1220

12. Salat-Canela C, Muñoz M, Sesé M et al (2015) Post-transcriptional regulation of connexins. Biochem Soc Trans 43:465–470

13. Maes M, Decrock E, Cogliati B et al (2014) Connexin and pannexin (hemi)channels in the liver. Front Physiol 4:405

14. Alexander DB, Goldberg GS (2003) Transfer of biologically important molecules between cells through gap junction channels. Curr Med Chem 10:2045–2058

15. Decrock E, Vinken M, de Vuyst E et al (2009) Connexin-related signaling in cell death: to live or let die? Cell Death Differ 16:524–536

16. Vinken M, Decrock E, De Vuyst E et al (2011) Connexins: sensors and regulators of cell cycling. Biochim Biophys Acta 1815:13–25

17. Chanson M, Derouette JP, Roth I et al (2005) Gap junctional communication in tissue inflammation and repair. Biochim Biophys Acta 1711: 197–207

18. Maes M, Cogliati B, Crespo YS et al (2015) Roles of connexins and pannexins in digestive homeostasis. Cell Mol Life Sci 72:2809–2821

19. King TJ, Bertram JS (2005) Connexins as targets for cancer chemoprevention and chemotherapy. Biochim Biophys Acta 1719:146–160

20. Kar R, Batra N, Riquelme MA et al (2012) Biological role of connexin intercellular channels and hemichannels. Arch Biochem Biophys 524:2–15

21. Jensen EC (2012) The basics of western blotting. Anat Rec 295:369–371

22. Kurien BT, Scofield RH (2006) Western blotting. Methods 38:283–293

23. Galva C, Gatto C, Milanick M (2012) Soymilk: an effective and inexpensive blocking agent for immunoblotting. Anal Biochem 426:22–23

24. Mruk DD, Cheng CY (2011) Enhanced chemiluminescence (ECL) for routine immunoblotting: an inexpensive alternative to commercially available kits. Spermatogenesis 1:121–122

25. Helenius A, McCaslin DR, Fries E et al (1979) Properties of detergents. Methods Enzymol 56:734–749

26. MacPhee DJ (2010) Methodological considerations for improving Western blot analysis. J Pharmacol Toxicol Methods 61:171–177

27. Olson BJ, Markwell J (2007) Assays for determination of protein concentration. Curr Protoc Protein Sci 48, Chapter 3:Unit 3.4

28. Noble JE, Bailey MJ (2009) Quantitation of protein. Methods Enzymol 463:73–95

29. Lovrien R, Matulis D (2005) Assays for total protein. Curr Protoc Microbiol Appendix 3:Appendix 3A

30. Smith PK, Krohn RI, Hermanson GT et al (1985) Measurement of protein using bicinchoninic acid. Anal Biochem 150:76–85

31. Pennisi M, Malaguarnera G, Puglisi V et al (2013) Neurotoxicity of acrylamide in exposed workers. Int J Environ Res Public Health 10: 3843–3854

32. Erdreich LS, Friedman MA (2004) Epidemiologic evidence for assessing the carcinogenicity of acrylamide. Regul Toxicol Pharmacol 39:150–157

33. Pruser KN, Flynn NE (2011) Acrylamide in health and disease. Front Biosci 3:41–51

34. Lipman NS, Jackson LR, Trudel LJ et al (2005) Monoclonal versus polyclonal antibodies: distinguishing characteristics, applications, and information resources. ILAR J 46:258–268

35. Silva JM, McMahon M (2014) The fastest Western in town: a contemporary twist on the classic Western blot analysis. J Vis Exp 84:e51149

36. Gilda JE, Gomes AV (2013) Stain-Free total protein staining is a superior loading control to β-actin for Western blots. Anal Biochem 440: 186–188

37. Alegria-Schaffer A (2014) Western blotting using chemiluminescent substrates. Methods Enzymol 541:251–259

38. Morseman JP, Moss MW, Zoha SJ et al (1999) PBXL-1: a new fluorochrome applied to detection of proteins on membranes. Biotechniques 26:559–563

39. Alegria-Schaffer A, Lodge A, Vattem K (2009) Performing and optimizing Western blots with an emphasis on chemiluminescent detection. Methods Enzymol 46:573–599

40. Wu M, Stockley PG, Martin WJ (2002) An improved western blotting technique effectively reduces background. Electrophoresis 23:2373–2376

41. Farmer SR, Wan KM, Ben-Ze'ev A et al (1983) Regulation of actin mRNA levels and translation responds to changes in cell configuration. Mol Cell Biol 3:182–189

42. Zhang JY, Zhang F, Hong CQ et al (2015) Critical protein GAPDH and its regulatory mechanisms in cancer cells. Cancer Biol Med 12:10–22

43. Bakhashab S, Lary S, Ahmed F et al (2014) Reference genes for expression studies in hypoxia and hyperglycemia models in human umbilical vein endothelial cells. G3 4: 2159–2165

44. Sangrajrang S, Denoulet P, Laing NM et al (1998) Association of estramustine resistance in human prostatic carcinoma cells with modified patterns of tubulin expression. Biochem Pharmacol 55:325–331

45. Prasad V, Kumar SS, Dey CS (2000) Resistance to arsenite modulates levels of alpha-tubulin and sensitivity to paclitaxel in Leishmania donovani. Parasitol Res 86:838–842

46. Maes M, McGill MR, da Silva TC et al (2016) Involvement of connexin43 in acetaminophen-induced liver injury. Biochim Biophys Acta in press

Chapter 4

Immunohisto- and Cytochemistry Analysis of Connexins

Bruno Cogliati, Michaël Maes, Isabel Veloso Alves Pereira, Joost Willebrords, Tereza Cristina Da Silva, Sara Crespo Yanguas, and Mathieu Vinken

Abstract

Immunohistochemistry (IHC) is a ubiquitous used technique to identify and analyze protein expression in the context of tissue and cell morphology. In the connexin research field, IHC is applied to identify the subcellular location of connexin proteins, as this can be directly linked to their functionality. The present chapter describes a protocol for fluorescent IHC to detect connexin proteins in tissues slices and cells, with slight modifications depending on the nature of biological sample, histological processing, and/or protein expression level. Basically, fluorescent IHC is a short, simple, and cost-effective technique, which allows the visualization of proteins based on fluorescent-labeled antibody–antigen recognition.

Key words Connexins, Antibody, Immunohistochemistry, Immunofluorescence, Immunocytochemistry, Microscopy, Morphology, Fluorophores, Localization

1 Introduction

Immunohistochemistry (IHC), or immunocytochemistry (ICC), is a commonly used technique to monitor protein expression and localization in tissue or in vitro cultures, respectively, using bright-field or fluorescent microscopy. Direct or indirect immunofluorescence is a powerful IHC-based technique that uses fluorescent-labeled antibodies to visualize protein expression while maintaining the composition, cellular characteristics, and structure of native tissue (Fig. 1) [1, 2]. Coon and colleagues were the first to describe the direct immunofluorescence technique using an antibody attached to a fluorescent dye, fluorescein isocyanate, to localize its respective antigen in a frozen tissue section [3, 4]. Subsequently, immunochemical methods based on peroxidase-labeled antibodies were introduced, allowing the development of new IHC, such as formalin-fixed paraffin-embedded (FFPE) tissues [5–9]. Currently, the use of antibodies to detect and localize individual or multiple

Mathieu Vinken and Scott R. Johnstone (eds.), *Gap Junction Protocols*, Methods in Molecular Biology, vol. 1437, DOI 10.1007/978-1-4939-3664-9_4, © Springer Science+Business Media New York 2016

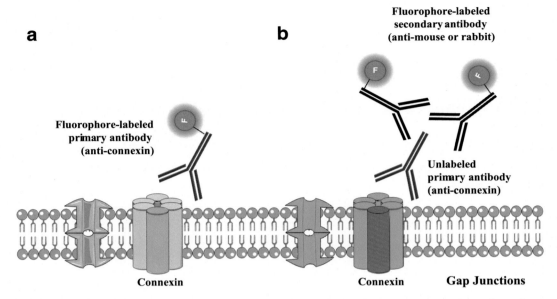

Fig. 1 *Direct and indirect immunofluorescence methods.* (**a**) Direct immunofluorescence method uses fluorophore-labeled primary antibody to bind directly to the connexin protein. This technique is rapid and quite specific. However, usually demands higher concentration of primary antibody and there are few options of antibodies conjugated directly to a fluorophore. (**b**) Indirect immunofluorescence method, or sandwich method, is a 2-step technique that uses an unlabeled primary antibody to bind the connexin protein, after which a fluorophore-labeled secondary antibody is used to detect the connexin antibody. This technique is more sensitive because more than one secondary antibody can bind to each primary antibody, which amplifies the fluorescence signal; however, this procedure has higher potential for cross-reactivity and immunostaining background, is more complicated and time consuming when compared to direct immunofluorescence

proteins in situ has developed into a powerful research tool in almost every field of biomedical research [10].

Gap junctions (GJs) are grouped in plaques at the plasma membrane surface of two adjacent cells and are composed of two juxtaposed connexons or hemichannels, each built up by six proteins named connexins [11]. At present, more than 20 connexin isotypes have been identified, which are expressed in a cell-specific way. Gap junction intercellular communication (GJIC) allows the direct flux of small and hydrophilic molecules, i.e., cyclic adenosine monophosphate (cAMP), inositol triphosphate (IP3), and ions, through GJs channels [12–15]. GJs are dynamic and the half-life cycle of connexins is short (<5 h) [16]. Connexins are biosynthesized on endoplasmic reticulum membranes and delivery happens to the plasma membrane as oligomerized hexameric hemichannels (connexons) [17]. Regulation of connexin synthesis can occur on transcriptional, translational, and posttranslational levels, resulting in a downregulation or lack of connexin expression and GJIC. In disease, connexin proteins can be abnormally localized within the cytoplasm. The exact mechanisms are still unknown, but impaired trafficking of the connexins to the membrane and increased internalization and degradation of connexons have been suggested [18–20]. It is known that alterations in the expression pattern and location of connexins are associated with potential oncogenesis and other chronic disorders, i.e., in liver and cardiac diseases [21–26]. In this regard, detection of aberrant subcellular location of connexin proteins is quite important to understand its role in pathological conditions.

In this chapter, fluorescent IHC-based protocols optimized to detect connexin proteins in cells or tissues slices will be outlined. Depending on the nature of biological sample, histological processing and/or protein expression level slight modifications are defined. The first step comprehends the adequate handling and fixation of cells or tissue specimens. The objective is to preserve tissue morphology and retain the antigenicity of the target proteins. To avoid loss during the procedure, cells or tissue sections should be placed on adhesive coating slides [1, 2]. For FFPE samples, tissue slides are deparaffinized with xylene and rehydrated in a series of ethanol solutions with decreasing concentrations. Afterward, the slides are subjected to heat-induced antigen retrieval (HIAR) in Tris–EDTA buffer (pH 9.0) or alternative method to reveal epitopes masked during the sample processing [27]. The background immunostaining caused by nonspecific antibody binding to endogenous Fc receptors or a combination of ionic and hydrophobic interactions should be blocked by bovine serum albumin (BSA), nonfat dry milk, gelatin, glycine, or normal serum from the species that the secondary antibody was raised in [28]. Incubation of monoclonal or polyclonal primary antibody is done for short (30–60 min, at 37 °C) or long time (overnight, at 4 °C) [1, 2]. Subsequently, the detection of connexins is performed using fluorescent-labeled secondary antibodies. This technique takes advantage of light emission with different spectral peaks against a dark background, with several options of fluorophores with different wavelengths of light emission (Table 1). The signal can be amplified by a tyramide signal amplification (TSA) method [28].

Table 1
List of most common dyes for immunofluorescence with their excitation and emission wavelength peaks

Color	Label (s)	Excitation (nm)	Emission (nm)
Blue	DAPI	365/50	450/65
Green	FITC; GFP; Alexa Fluor® 488	475/40	535/45
Yellow	TRITC; Alexa Fluor® 555, 546, 568; Cy3	546/12	585/40
Red	Alexa Fluor® 594; Propidium iodide; eFluor® 615; Texas Red	560/55	645/75
Near infrared	eFluor® 660; Alexa Fluor® 647; Cy5	620/60	700/75
Infrared	IRDye 800®	787	812

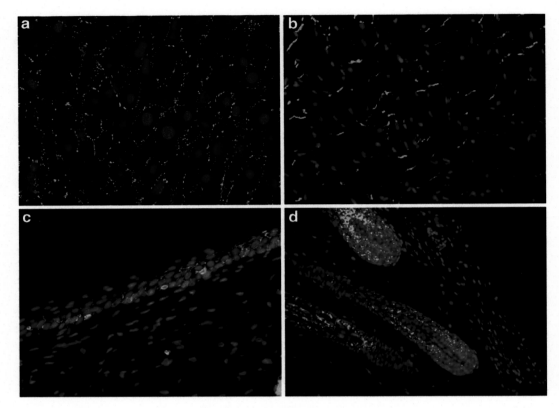

Fig. 2 *Detection of connexins (Cx) in frozen and paraffin-embedded tissue sections.* (**a**) Immunodetection of Cx32 in frozen sections of human liver tissue fixed in ice cold acetone, using 1/500 anti-Cx32 polyclonal antibody produced in rabbit (Sigma-Aldrich, USA), 1/200 Alexa Fluor® 488-conjugated secondary antibody polyclonal goat anti-rabbit IgG (H+L) (Life Technologies, USA), and 1/1000 DAPI (Life Technologies, USA) for nuclear counter stain (200×). (**b**) Cx43 detected in paraffin-embedded sections of mouse heart fixed in methacarn solution, using 1/500 anti-Cx43 polyclonal antibody produced in rabbit (Sigma-Aldrich, USA), 1/200 Alexa Fluor® 488-conjugated secondary antibody polyclonal goat anti-rabbit IgG (H+L) (Life Technologies, USA) and 1/1000 propidium iodide (Sigma-Aldrich, USA) for nuclear counter stain (400×). (**c, d**) Cx43 detected in paraffin-embedded sections of mouse skin epidermis and hair bulb using the tyramide signal amplification (TSA) method with 1/1000 anti-Cx43, polyclonal antibody produced in rabbit (Sigma-Aldrich, USA), 1/500 polyclonal goat anti-rabbit Immunoglobulins/HRP (Dako Cytomatic, USA), 1/100 tyramide working solution and 1/1000 propidium iodide (Sigma-Aldrich, USA) for nuclear counter stain (400×)

Finally, the slides are incubated with a DNA-fluorescent marker for the nuclei counterstain, mounted on antifade media to avoid photobleaching and then visualized under a fluorescent microscope with the appropriate filters (Fig. 2).

2   Materials

2.1 Frozen Tissue Sections

1. Optimal cutting temperature (OCT)-embedding medium (Leica Biosystems, Germany).

2. Cryomolds.

3. Liquid nitrogen.

4. Isopentane.

5. Cryostat.

6. Disposable blades and brushes.

7. Microscope adhesive glass slides (25 × 75 mm) (*see* **Note 1**).

8. Acetone, histological grade.

2.2 Paraffin-Embedded Tissue Sections

1. Methacarn solution. Methanol:chloroform:glacial acetic (60:30:10 v/v). This solution can be stored for 3 months at 4 °C.

2. 10 % neutral buffered Formalin: 4 g NaH_2PO_4, 6.5 g Na_2HPO_4, 100 mL formaldehyde 37–40 % (w/v) (*see* **Note 2**), 900 mL distilled water. Mix and adjust for pH 6.8. This solution can be stored at room temperature.

3. Xylene, histological grade (*see* **Note 3**).

4. Xylene: 100 % Ethanol (1:1, v/v).

5. Ethanol, anhydrous denatured, histological grade (70, 95 and 100 % EtOH). Prepare the EtOH solutions in distillated water (v/v).

6. Fume hood.

7. Microtome, blades, and thermostatic water bath.

8. Microscope adhesive glass slides (25 × 75 mm).

9. Oven (55–65 °C).

10. Tris-ethylenediaminetetraacetic acid (EDTA) buffer: 10 mM Tris Base, 1 mM EDTA Solution, 0.05 % Tween® 20 in ultrapure water. Adjust to pH 9.0. This solution can be stored at room temperature for 3 months or at 4 °C for longer storage.

11. Electronic pressure cook (*see* **Note 4**).

2.3 In Vitro Cultures

1. Cell line or primary cells.

2. Cell culture medium and supplements.

3. Thermostatic water bath (37 °C).

4. Laminar flow cabinet.

5. CO_2 humidified incubator (37 °C ± 1 °C, 90 % ± 5 % humidity, 5 % ± 1 % CO_2).

6. Tissue culture plates (6-, 12-, or 24-well plates) or chamber culture slides (*see* **Note 5**).

7. Sterile glass coverslips (*see* **Note 6**).

8. Acetone, histological grade.

2.4 General Procedure for Fluorescent IHC

1. Vessels with slide rack and plastic slide holders.

2. Humidified chamber (*see* **Note 7**).

3. Shaker.

4. Refrigerator or refrigerated incubator (2–8 °C).

5. Wash buffer: 10× Phosphate-buffered saline (PBS) with Tween® 20. 70 mM Na_2HPO_4, 30 mM NaH_2PO_4, 1.37 M NaCl, 0.5 % Tween® 20 in distilled water (*see* **Note 8**). Adjust to pH 7.2–7.4 (*see* **Note 9**). To prepare 1 L of 1× PBS: add 100 mL of 10× PBS with Tween® 20–900 mL of distilled water. Mix and adjust pH if necessary. This solution can be stored up to 6 months at room temperature.

6. Blocking buffer: 5 % (m/v) nonfat dry milk powder in 1× PBS (*see* **Note 10**). Prepare *ex tempore*.

7. Buffer for antibody dilution: 1 % BSA (Sigma-Aldrich, USA) and 0.1 % Sodium Azide diluted in 1× PBS without Tween® 20. This solution can be stored for 3 months at 4 °C (*see* **Note 11**).

8. Appropriate monoclonal or polyclonal primary antibodies (*see* **Note 12**).

9. Isotype IgG or preimmune sera (*see* **Note 13**).

10. Alexa Fluor® 488 goat anti-mouse or Alexa Fluor® 594 goat anti-rabbit as appropriate secondary antibody (Life Technologies, USA). Other labels can also be used (i.e., FITC or Texas Red) (Table 1).

11. Nuclear dyes: propidium iodide (PI) (Sigma-Aldrich, USA) diluted in 1× PBS without Tween® 20 (1 µg/mL) or 4′,6-diamidino-2-phenylindole (DAPI) (Life Technologies, USA) diluted in 1× PBS without Tween® 20 (10 µg/mL).

12. Aqueous antifading mounting medium (Vector Laboratories, USA).

13. Glass coverslips.

14. Nail polish.

15. Fluorescence microscope with the appropriated filters (Table 1).

2.5 Tyramide Signal Amplification (TSA)

1. TSA Fluorescein Kit (Perkin Elmer, USA).

2. Blocking buffer: 0.1 M Tris–HCl (pH 7.5), 0.15 M NaCl, 0.5 % Blocking Reagent. Prepare according to the manufacturer's recommendations (Perkin Elmer, USA).

3. Fluorophore tyramide stock and working solution. Prepare according to the manufacturer's recommendations (Perkin Elmer, USA).

4. Polyclonal Goat Anti-Rabbit or Mouse Immunoglobulins/Horseradish peroxidase (HRP) (Dako Cytomatic, USA).

3 Methods

3.1 Fluorescent IHC in Frozen Tissue Sections

1. Snap freeze fresh tissue in liquid nitrogen or isopentane precooled in liquid nitrogen (*see* **Note 14**). Store flash frozen tissue at −80 °C.

2. Embed flash-frozen tissue in OCT compound in cryomolds. This can be stored at –80 °C.

3. Cut 5–10 μm thick sections in the cryostat and mount on microscope adhesive glass slides (*see* **Note 15**).

4. Store slides at –80 °C for long-term storage or at –20 °C for few weeks. For best results, use them immediately.

5. Before staining, fix the slides in ice cold acetone for 10–20 min at –20 °C (*see* **Note 16**).

6. Wash slides three times for 5 min with wash buffer on a shaker.

7. Incubate slides with monoclonal or polyclonal primary antibody diluted in antibody dilution buffer for 1 h at 37 °C in a humidified chamber (*see* **Note 17**). Use the appropriate negative and positive controls (*see* **Note 18**).

8. Wash slides three times for 5 min with wash buffer on a shaker.

9. Incubate slides with a secondary antibody Alexa Fluor® 488 goat anti-mouse or Alexa Fluor® 594 goat anti-rabbit, diluted in antibody dilution buffer (10 μg/mL), for 1 h at room temperature in a humidified chamber. From this point, the procedures should be done in the dark to avoid photobleaching.

10. Wash three times for 5 min with wash buffer on a shaker.

11. Perform the nuclear counter stain with PI for 15 min or with DAPI for 5 min at room temperature (*see* **Note 19**).

12. Wash three times for 5 min with wash buffer on a shaker.

13. Mount coverslip with a drop of antifading mounting medium.

14. Seal coverslip with nail polish (*see* **Note 20**).

15. Store in dark at 4 °C.

3.2 Fluorescent IHC in Paraffin-Embedded Tissue Sections

1. Cut the tissue specimens in small fragments (*see* **Note 21**) and immerse in methacarn fixative solution for 12 h (*see* **Note 22**) or in 10 % neutral buffered formalin for 18–24 h (*see* **Note 23**).

2. Embedded fixed samples in paraffin wax (Table 2).

3. Cut 3–5 μm tick sections using a rotary microtome and float the sections in 40–45 °C thermostatic water bath.

4. Place the sections on microscope adhesive glass slides within the water bath and remove the water excess.

5. Dry slices in an incubator at 56–60 °C for 2 h (*see* **Note 24**).

6. Store slides at 4 °C for short period and at –20 °C or –80 °C for long-term usage (*see* **Note 25**).

7. Before proceeding with the staining protocol, the slides should be deparaffined in xylene (*see* **Note 26**) and rehydrated in ethanol (Table 3).

8. Place slides in plastic slide holders and fill with Tris–EDTA buffer.

Table 2
Tissue processing for *paraffin wax embedding*

Solution	Incubation/time
70% Ethanol	Two changes[a], 1 h each
80% Ethanol	One change, 1 h
95% Ethanol	One change, 1 h
100% Ethanol	Three changes[a], 1.5 h each
Xylene	Three changes[a], 1.5 h each
Paraffin wax (58–60 °C)	Two changes[a], 2 h each

[a]Use different solutions for each change

Table 3
Deparaffin and rehydration processing of paraffin-embedded sections

Solution	Incubation/time
Xylene	Two changes[a], 30 min each
Xylene:100% Etanol (1:1)	One change, 30 min
100% Ethanol	Two changes[a], 5 min each
95% Etanol	One change, 5 min
70% Etanol	One change, 5 min
Running tap water to rinse	10 min
Distillated water on a shaker	10 min

[a]Use different solutions for each change

9. Place holders in an electric pressure cooker and add at least 1.5 L of distillate water to pressure cooker chamber.

10. Set target temperature and heating time according to the manufacturer's instructions, normally boil for 30 s to 5 min and allow the pressure cooker to cool for 20 min prior to opening (*see* **Note 27**).

11. Immediately after, rinse the slides in running tap water for 10 min

12. Wash slides in distillate water for 10 min on a shaker.

13. Place slides in wash buffer for 5 min on a shaker.

14. If the signal of a specific connexin needs to be amplified, perform Tyramide signal amplification (*see* **Note 28**).

15. Incubate slides with primary antibody diluted in antibody dilution buffer (*see* **Note 17**) overnight at 4 °C in a humidified

chamber (*see* **Note 29**). Use the appropriate negative and positive controls (*see* **Note 18**).

16. Wash slides three times for 5 min with wash buffer on a shaker.

17. Incubate slides with a secondary antibody Alexa Fluor® 488 goat anti-mouse or Alexa Fluor® 594 goat anti-rabbit (10 μg/mL), diluted in antibody dilution buffer, for 1 h at room temperature in a humidified chamber. From this point, the procedures should be done in the dark to avoid photobleaching.

18. Wash slides three times for 5 min with wash buffer on a shaker.

19. Perform the nuclear counter stain with PI for 15 min or with DAPI for 5 min at room temperature (*see* **Note 19**).

20. Wash slides three times for 5 min with wash buffer on a shaker.

21. Mount coverslip with a drop of antifading mounting medium.

22. Seal coverslip with nail polish (*see* **Note 20**).

23. Store in dark at 4 °C.

3.3 Fluorescent Immunocytochemistry

1. Grow cultured cells on sterile glass coverslips or alternatively culture cells in 1 to 8-chamber glass slides (*see* **Note 30**).

2. Aspirate the supernatant from each chamber/well and rinse the cells three times for 5 min with ice cold wash buffer on a shaker.

3. Fix cells using ice cold acetone for 10 min at –20 °C. When removing acetone make sure the cells are not allowed to dry out (*see* **Note 31**).

4. Wash cells three times for 5 min with wash buffer.

5. Incubate cells with 5 % skim milk diluted in PBS for 30 min at room temperature.

6. Wash cells three times for 5 min with wash buffer.

7. Incubate cells with primary antibody diluted in antibody dilution buffer for 1 h at 37 °C or overnight at 4 °C in a humidified chamber (*see* **Note 17**). Use the appropriate negative and positive controls (*see* **Note 18**).

8. Wash three times for 5 min with wash buffer on a shaker.

9. Incubate slides with a secondary antibody Alexa Fluor® 488 goat anti-mouse or Alexa Fluor® 594 goat anti-rabbit (10 μg/mL), diluted in antibody dilution buffer, for 1 h at room temperature in a humidified chamber. From this point, the procedures should be done in the dark to avoid photobleaching.

10. Wash three times for 5 min with wash buffer on a shaker.

11. Perform the nuclear counter stain with PI for 15 min or with DAPI for 5 min at room temperature (*see* **Note 19**).

12. Wash slides three times for 5 min with wash buffer on a shaker.

13. Mount coverslip with a drop of antifading mounting medium on a microscope glass.

14. Seal coverslip with nail polish (*see* **Note 20**).

15. Store in dark at 4 °C.

3.4 Tyramide Signal Amplification (TSA) Method

1. Quench endogenous peroxidase activity by incubating slides in 6% H_2O_2 in 1× PBS for 30 min in dark (*see* **Note 32**).

2. Wash slides three times for 5 min with wash buffer on a shaker.

3. Block slides for 30 min in blocking buffer at room temperature.

4. Incubate slides with primary antibody diluted in blocking buffer overnight at 4 °C in a humidified chamber (*see* **Note 17**). Use the appropriate negative and positive controls (*see* **Note 18**).

5. Wash slides three times for 5 min with wash buffer on a shaker.

6. Incubate slides with a polyclonal goat anti-rabbit or mouse immunoglobulins/HRP, diluted in blocking buffer (1:500) for 30 min at room temperature.

7. Wash slides three times for 5 min with wash buffer on a shaker.

8. Incubate in tyramide working solution (1:100) for 10 min at room temperature. From this point, the procedures should be done in the dark to avoid photobleaching.

9. Wash slides three times for 5 min with wash buffer on a shaker.

10. Perform the nuclear counter stain with PI for 15 min or with DAPI for 5 min at room temperature (*see* **Note 19**).

11. Wash slides three times for 5 min with wash buffer on a shaker.

12. Mount coverslip with a drop of antifading mounting medium.

13. Seal coverslip with nail polish (*see* **Note 20**).

14. Store in dark at 4 °C.

4 Notes

1. Use of an adhesive coating, such as poly-l-lysine, charging, or silanization, can improve the tissue adherence and reduce damages in fragile tissues (i.e., dermis skin layer or adipose tissue) or to avoid tissue loss during the immunostaining procedure. This is strongly recommended when enzymatic or heat-induced antigen retrieval methods are applied in frozen or paraffin-embedded tissue slides [1, 2].

2. Formalin is formaldehyde gas dissolved in water and reaches saturation at 37–40% formaldehyde (this solution is considered as 100% formalin). In this context, 10% formalin actually represents 10% of a 37–40% stock solution; this means that the actual amount of dissolved formaldehyde in the 10% formalin is therefore only 3.7–4.0%. As formaldehyde is a severe eye and

skin irritant with known carcinogenic and corrosive potential, always work in well-ventilated area and wear goggles, gloves, and lab coat [28, 29].

3. Xylene is an aromatic hydrocarbon and is potentially dangerous due to its volatility and inflammability capacities. Acute and chronic exposure to xylene is associated with nonspecific clinical signals related to the central nervous system. It must be used in a fume hood to avoid occupational hazard [30].

4. Heat-induced antigen retrieval by microwave radiation may lead to inconsistent IHC results. Because pressure cookers or vegetable steamers induce uniform and gentle heat, they are currently used in most laboratories.

5. Choice of slide design is often dictated by the experiment. Some slides have four wells, others have eight some are from glass, and others are from plastic. Glass is recommended as the slide becomes more versatile, acetone can be used as a fixative and slides can be dehydrated in ethanol and cleared in xylene. However, some cell types need a plastic surface to grow with a good cell confluence, i.e., rat primary hepatocytes.

6. 12 mm circular glass coverslips fit in 24-well plates. These coverslips may be autoclaved in a petri dish. Most cells adhere well to glass. Cells that do not adhere may require treatment of the glass with an adhesive coating, such as poly-l-lysine, silane, or other attachment factor.

7. The humidified chamber can be either the incubator itself or a container with wet paper towels. This procedure is important to prevent evaporation and drying of the tissue sections during the incubation time as this may cause nonspecific antibody binding and therefore high background staining.

8. The use of Tween® 20 in the wash buffer helps to reduce surface tension, allowing reagents to cover the whole tissue section with ease. It is also believed to dissolve Fc receptors, therefore reducing nonspecific binding.

9. Do not adjust pH using concentrated HCl or NaOH. If necessary, use a separate 0.2 M solution of either the monobasic or dibasic sodium phosphate (depending on how you need to adjust the pH) and add accordingly.

10. When preparing blocking buffer, make sure that all milk powder is dissolved by stirring the solution for about 20 min. It may be necessary to heat the solution slightly to fully dissolve nonfat dry milk. This solution can be used up to 5 days when stored at 4 °C. It is recommended to mix before use. Alternative block solutions are 1% gelatin or 10% serum from the species from which the secondary antibody was raised in.

11. Do not use this buffer to dilute HRP conjugated antibody as the sodium azide is inhibitor of HRP.

12. Monoclonal antibodies have high affinity and specificity to a single epitope, which can be directly affected by structural changes in the proteins during the fixation and histological processing. On the other hand, polyclonal antibodies are more stable and can recognize multiple epitopes; this last one represents the main reason why this kind of antibody is more frequently used for IHC. However, polyclonal antibodies can induce higher background immunostaining and nonspecific bindings should be efficiently blocked [1, 2, 27].

13. The isotype IgG control should be applied when working with monoclonal primary antibodies as a negative control test. The slides are incubated with a nonimmune immunoglobulin of the same isotype and concentration as the primary monoclonal antibody. For polyclonal antibodies, the slides should be incubated with preimmune sera. In addition, cells and tissue from genetically engineered animals, as well as siRNA, with PCR confirmation, can be used as positive and negative controls [31].

14. Snap freezing is the process by which samples reach ultralow temperatures very rapidly using nitrogen liquid or isopentane cooled in nitrogen liquid. In contrast to the slow freezing, this method reduces the ice crystal formation which causes distortion in cell morphology and, consequently, a low-quality cryosection. Take serious care using nitrogen liquid and use all equipments for individual security.

15. Frozen tissue sections give much better antigen preservation than paraffin-embedded tissue sections. However, the morphological details and resolution are usually reduced. The temperature in the cryostat must be correct for the specimen being cut. The microtome and the antiroll plate must be correctly adjusted and operated. The cutting blade must be sharp [32].

16. Air-dried frozen tissues are not recommended for IHC; it is better to choose and test different fixative for each kind of protein [28]. In our experience, tissue slices fixated with acetone show satisfactory results for connexins immunostaining.

17. The primary antibody should be diluted to the manufacturer's recommendations or to a previously optimized dilution. A plethora of companies have primary connexin antibodies available, including Sigma Aldrich, Abcam, Thermo Fischer, Santa Cruz, EMD Millipore, Genetex, ProteinTech, Alomone, Biocompare, Life Technologies, and Merck Millipore. It is recommended to test different concentrations of the primary antibody and keep the concentration of the secondary antibody low to avoid nonspecific binding (Fig. 2). Most antibodies will be used in IHC at a concentration between 0.5 and 10 µg/mL [1, 2, 28].

18. Besides the negative control with isotype IgG or preimmune sera, it is strongly recommended to include a positive control

to ensure that the antibody is performing as expected, i.e., liver and heart cells or tissue for Cx32 and Cx43 immunostaining, respectively.

19. Both the excitation and emission wavelengths are specific characteristics for each fluorophore; however, they may show overlapping fluorescence emission wavelengths among different fluorescent labels and nuclear dyes, i.e., Alexa Fluor® 594 is visually similar in color with the PI nuclear dye (Table 1).

20. Sealing the coverslips with nail polish prevents drying and movement under microscopy.

21. To enhance penetration of the fixative during immersion fixation, it is recommended that tissues be no thicker than 10 mm. For complete fixation, the ideal volume of formaldehyde solution compared to tissue volume should be in the ratio of 1:25 (w/v), with a minimum ratio of 1:10 (w/v) [29].

22. Methacarn is a non-cross-linking protein-precipitating fixative that was shown to maintain tissue morphology and usually give superior immunohistochemical results than aldehyde-based cross-linking fixatives, because antigenicity is usually maintained intact [33]. In our experience, methacarn is the best option of fixative solution to immunostaining connexins in paraffin-embedded tissues.

23. Prolonged fixation in formalin (more than 24 h) results in gradual loss of the antigenicity, which will require heat-induced antigen retrieval methods to unmask the epitopes and allow the antigen–antibody binding. It is very important to optimize fixation conditions since underfixation or overfixation may reduce or abolish tissue antigenicity [29].

24. This also will allow a better slice adherence in the slides and remove the paraffin excess from the tissue section. Do not allow higher temperature and time incubation, as this may reduce or abolish tissue antigenicity.

25. Storage of unstained tissue sections longer than 2 months can promote loss of antigenicity by oxidation, protein degradation by high temperature, or other unknown factors. Then, it is recommended to use fresh cut sections or storage of the unstained sections paraffin coated in vacuum-sealed desiccators and cold temperature (at 4 °C, –20 °C or –80 °C) [10].

26. Due to the potential for autofluorescence with FFPE tissues, the first xylene can be used at 56–60 °C for 30 min. Additionally, the deparaffinized section can be incubated with 1 mmol/L glycine in 1× PBS for 30 min at room temperature and, after washes, with 1 mg/mL $NaBH_4$ in 1× PBS to further reduce autofluorescence in FFPE slides [34].

27. Pretreatment with these solutions may induce a dramatic enhancement of immunoreactivity. However, they also have

the potential to affect tissue morphology. The optimal method of antigen retrieval must be determined experimentally, using different incubation time, temperatures, and alternative options of buffer, i.e., citrate buffer (pH 6.0) [27]. Proteolytic enzymes, i.e., Proteinase K and Trypsin, also can be tested in FFPE slides. The concentration of enzyme and incubation time will depend on type of tissue and fixation but is usually 0.05–0.1% for 5–30 min at 37 °C [2].

28. The TSA method is based on the ability of HRP to catalyze in the presence of hydrogen peroxide the oxidation of the phenol moiety of labeled tyramide onto protein surrounding the HRP. This allows an increase in the detection of an antigenic site up to 100-fold compared to the conventional indirect method, with no loss in resolution [1, 2]. In this context, the TSA detection method is recommended to identify connexins with low expression levels.

29. Overnight incubation allows the use of lower titer of antibodies and reduces nonspecific background staining.

30. Grow the cells at a concentration that will allow the cells to spread out without growing on top of each other (around 70–80% of confluence).

31. If the target protein is localized intracellularly, it is very important to permeabilize the cells. Acetone or methanol-fixed samples do not require permeabilization. If cells were fixed in 4% paraformaldehyde or alternative fixative, permeabilize cells in acetone or 100% ethanol for 10 min at −20 °C. Alternatively, use 100 μM Digitonin or 0.5% Saponin diluted in 1× PBS for 10 min at room temperature. Cell permeabilization with Triton X-100 is not recommended for membrane-associated antigens (i.e., connexins) since it destroys membranes.

32. The blockage of endogenous peroxidase is necessary as the activation and covalent binding of TSA reagent is catalyzed by peroxidase. The incubation period can range from 10 to 60 min, depending on the endogenous activity of peroxidase in each kind of tissue.

Acknowledgements

This work was financially supported by the University Hospital of the Vrije Universiteit Brussel-Belgium (Willy Gepts Fonds UZ-VUB), the Fund for Scientific Research-Flanders (FWO grants G009514N and G010214N), the European Research Council (ERC Starting Grant 335476), the University of São Paulo-Brazil and the São Paulo Research Foundation (FAPESP SPEC grant 2013/50420-6).

References

1. Taylor CR, Shi S-R, Barr NJ (2011) Techniques of immunohistochemistry: principles, pitfalls, and standardization. In: Dabbs DJ (ed) Diagnostic immunohistochemistry, 3rd edn. W.B. Saunders, Philadelphia, pp 1–41

2. Buchwalow IB, Böcker W (2010) Immunohistochemistry: basics and methods. Springer, Germany

3. Coons AH, Creech HJ, Jones RN (1941) Immunological properties of an antibody containing a fluorescent group. Exp Biol Med 47:200–202

4. Coons AH, Creech HJ, Jones RN et al (1942) The demonstration of pneumococcal antigen in tissues by the use of fluorescent antibody. J Immunol 45:159–170

5. Nakane PK, Pierce GB (1967) Enzyme-labeled antibodies for the light and electron microscopic localization of tissue antigens. J Cell Biol 33:307–318

6. Avrameas S (1972) Enzyme markers: their linkage with proteins and use in immunohistochemistry. Histochem J 4:321–330

7. Taylor CR, Burns J (1974) The demonstration of plasma cells and other immunoglobulin-containing cells in formalin-fixed, paraffin-embedded tissues using peroxidase-labelled antibody. J Clin Pathol 27:14–20

8. Taylor CR (1994) The current role of immunohistochemistry in diagnostic pathology. Arch Pathol Lab Med 7:59–105

9. Taylor CR (1994) An exaltation of experts: concerted efforts in the standardization of immunohistochemistry. Hum Pathol 25:2–11

10. O'Hurley G, Sjöstedt E, Rahman A et al (2014) Garbage in, garbage out: a critical evaluation of strategies used for validation of immunohistochemical biomarkers. Mol Oncol 8:783–798

11. Goodenough DA, Goliger JA, Paul DL (1996) Connexins, connexons, and intercellular communication. Annu Rev Biochem 65:475–502

12. Yamasaki H, Naus CC (1996) Role of connexin genes in growth control. Carcinogenesis 17:1199–1213

13. Bruzzone R, White TW, Paul DL (1996) Connections with connexins: the molecular basis of direct intercellular signaling. Eur J Biochem 238:1–27

14. King TJ, Bertram JS (2005) Connexins as targets for cancer chemoprevention and chemotherapy. Biochim Biophys Acta 1719:146–160

15. Evert M, Ott T, Temme A et al (2002) Morphology and morphometric investigation of hepatocellular preneoplastic lesions and neoplasms in connexin32-deficient mice. Carcinogenesis 23:697–703

16. Musil LS, Le AC, VanSlyke JK et al (2000) Regulation of connexin degradation as a mechanism to increase gap junction assembly and function. J Biol Chem 275:25207–25215

17. Krutovskikh VA, Mesnil M, Mazzoleni G et al (1995) Inhibition of rat liver gap junction intercellular communication by tumor-promoting agents in vivo. Association with aberrant localization of connexin proteins. Lab Invest 72:571–577

18. Rose B, Mehta PP, Loewenstein WR (1993) Gap-junction protein gene suppresses tumorigenicity. Carcinogenesis 14:1073–1075

19. Trosko JE, Chang CC, Madhukar BV et al (1990) Chemical, oncogene and growth factor inhibition gap junctional intercellular communication: an integrative hypothesis of carcinogenesis. Pathobiology 58:265–278

20. Yamasaki H (1990) Gap junctional intercellular communication and carcinogenesis. Carcinogenesis 11:1051–1058

21. Agullo-Pascual E, Delmar M (2012) The non-canonical functions of Cx43 in the heart. J Membr Biol 245:477–482

22. Nakashima Y, Ono T, Yamanoi A et al (2004) Expression of gap junction protein connexin32 in chronic hepatitis, liver cirrhosis, and hepatocellular carcinoma. J Gastroenterol 39:763–768

23. Nakata Y, Iwai M, Kimura S et al (1996) Prolonged decrease in hepatic connexin32 in chronic liver injury induced by carbon tetrachloride in rats. J Hepatol 25:529–537

24. Duffy HS (2012) The molecular mechanisms of gap junction remodeling. Heart Rhythm 9:1331–1334

25. Kieken F, Mutsaers N, Dolmatova E et al (2009) Structural and molecular mechanisms of gap junction remodeling in epicardial border zone myocytes following myocardial infarction. Circ Res 104:1103–1112

26. Martins-Marques T, Catarino S, Marques C et al (2015) Heart ischemia results in connexin43 ubiquitination localized at the intercalated discs. Biochimie 112:196–201

27. Hayat MA (2002) Microscopy, immunohistochemistry, and antigen retrieval methods: for light and electron microscopy. Springer, New York

28. Kumar GL, Rudbeck L (2009) Education guide: immunohistochemical staining methods: pathology. Dako, North America

29. Thavarajah R, Mudimbaimannar VK, Elizabeth J et al (2012) Chemical and physical basics of routine formaldehyde fixation. J Oral Maxillofac Pathol 16:400–405

30. Kandyala R, Raghavendra SP, Rajasekharan ST (2010) Xylene: an overview of its health hazards and preventive measures. J Oral Maxillofac Pathol 14:1–5

31. Hewitt SM, Baskin DG, Frevert CW et al (2014) Controls for immunohistochemistry: the Histochemical Society's standards of practice for validation of immunohistochem-ical assays. J Histochem Cytochem 62: 693–697

32. Peters SR (2010) A practical guide to frozen section technique. Springer, New York

33. Mitchell D, Ibrahim S, Gusterson BA (1985) Improved immunohistochemical localization of tissue antigens using modified methacarn fixation. J Histochem Cytochem 33:491–495

34. Koval M, Billaud M, Straub AC et al (2011) Spontaneous lung dysfunction and fibrosis in mice lacking connexin 40 and endothelial cell connexin 43. Am J Pathol 178:2536–2546

Chapter 5

Small Interfering RNA-Mediated Connexin Gene Knockdown in Vascular Endothelial and Smooth Muscle Cells

Miranda E. Good, Daniela Begandt, Leon J. DeLalio, Scott R. Johnstone, and Brant E. Isakson

Abstract

Global knockout of vascular connexins can result in premature/neonatal death, severe developmental complications, or compensatory up-regulation of different connexin isoforms. Thus, specific connexin gene knockdown using RNAi-mediated technologies is a technique that allows investigators to efficiently monitor silencing effects of single or multiple connexin gene products. The present chapter describes the transient knockdown of connexins in vitro and ex vivo for cells of the blood vessel wall. In detail, different transfection methods for primary endothelial cells and ex vivo thoracodorsal arteries are described. Essential controls for validating transfection efficiency as well as targeted gene knockdown are explained. These protocols provide researchers with the ability to modify connexin gene expression levels in a multitude of experimental setups.

Key words Transfection, Lipofection, Electroporation, siRNA, Off-target, Knockdown, Ex vivo transfection

1 Introduction

RNA interference (RNAi) is a naturally occurring mechanism that was first described in *Caenorhabditis elegans* [1, 2]. Since then, studies about the mechanisms and molecular machinery have been carried out in a number of model organisms [3, 4]. There are two main RNAi pathways that differ in the origin of interfering RNA molecules and the mechanisms of action; however, both mechanisms produce similar outcomes. The basic underlying mechanism is derived from a host defense mechanism that exists to prevent foreign RNA viruses from replicating inside the host cells [5]. Host cells can recognize the viral double-stranded RNA (dsRNA) by Dicer, a cytoplasmic ribonuclease, which cuts dsRNA into small interfering RNAs (siRNAs) of 21–25 bp in length to prevent the

Mathieu Vinken and Scott R. Johnstone (eds.), *Gap Junction Protocols*, Methods in Molecular Biology, vol. 1437,
DOI 10.1007/978-1-4939-3664-9_5, © Springer Science+Business Media New York 2016

translation of viral RNA into protein [6–8]. Another mechanism of RNA interference is the internal regulation of gene expression by another class of siRNAs, the micro RNAs (miRNAs) with 19–25 bp in length. miRNAs are encoded in the genome, synthesized in a multistep process, and cleaved into active forms of RNA by Dicer [9, 10]. Both types of interfering RNAs (i.e., siRNA and miRNA) induce the formation of the RNA-induced silencing complex (RISC) in the cytoplasm. The RISC complex binds the double-stranded siRNA/miRNA, unwinds the RNA, and binds the guide strand, which is complementary to the targeted mRNA. In a subsequent step, the RISC complex binds targeted mRNA, which leads to the degradation, cleavage, or blocking of mRNA preventing the translation of mRNA to protein [11–13]. One main difference between the two different interfering RNAs is that siRNAs have complete sequence specificity compared to miRNAs and results in the physical cleavage of the siRNA-targeted mRNA [14–16].

To induce targeted knockdown of a gene of interest (GOI), researchers can use the RNAi mechanism by exogenously applying interfering RNAs to induce the sequence-specific blockade or degradation of mRNA with subsequent translational silencing [17, 18]. In general, there are three different ways to induce RNAi: (1) commercially available double-stranded siRNAs designed with perfect complementarity and unique sequences to the targeted mRNA that have to be exogenously applied and introduced into the cell; (2) plasmids encoding short hairpin RNAs (shRNAs), which are the preprocessed form of siRNAs, can also be used to induce the RNAi effect; or (3) viral-based vectors that express shRNAs can be introduced into difficult to transfect cells or tissues and can provide long-term gene silencing [1, 19, 20]. These given RNAi-inducing systems must be carefully selected to match the appropriate silencing method with the experimental specimen type.

Ultimately, the success of RNAi experiments depends on the specificity and efficiency of the siRNA. Different companies such as GE Healthcare Dharmacon, Inc. or Ambion Thermo Fisher Scientific® have developed specific design algorithms to identify potent siRNA sequences to offer a library of well-validated predesigned siRNAs. However, highly complementarity siRNA sequences can still induce off-target effects [21, 22] (*see* **Note 1**). Pooling of several independent siRNAs that target the same gene is a common strategy that has been shown to yield efficient gene silencing while reducing the frequency of off-target effects [23]. Furthermore, transfection efficiency and cell survival is another important step for a successful RNAi gene expression regulation.

Cell lines, primary cells, and tissue differ in their ability to be transfected. Here we present a classical transfection protocol for primary endothelial cells, which can be adjusted to a number of different cell types. We also present a modified siRNA-mediated gene silencing protocol for ex vivo endothelial and smooth muscle targeted gene knockdown of intact blood vessels [24]. These two different methods allow the researcher to adjust the experimental setup to the investigated cell type, cell line, or tissue and to efficiently silence their GOI. Although outside the scope of these protocols, it should be noted that deep-tissue applications or long-term RNAi-based gene silencing strategies should be adjusted to the experimental setup by the use of laser- or viral-based transfection systems [25–28].

Often multiple isoforms from a family of genes are expressed within the same cell resulting in difficulty in isolating the unique contributions of each protein to the function of the cell. Such is the case with connexins, which also have been shown to change their expression patterns in isolated primary cells. Knockdown of a specific or multiple connexin(s) allows for isolated functional analysis for each isoform [29, 30]. Thus, the current chapter describes a mechanism for knocking down GOIs in both primary cells and ex vivo blood vessels.

2 Materials

2.1 Cell Culture of Primary Vascular Endothelial Cells

1. Human umbilical endothelial cells (HUVEC) purchased at passage 2 (*see* **Note 2**).

2. Endothelial cell medium: Medium 200 supplemented with low serum growth kit (Thermo Fisher Scientific) and 20% FBS (*see* **Note 3**).

3. Incubator at 37 °C, 5% CO_2, 90% humidity, and class 2 biosafety cabinet.

4. Sterile cell culture 6- or 12-well plastic plates.

5. 50 μM fibronectin solution (*see* **Note 4**).

6. Plastic cell scraper.

7. Trypsin–EDTA.

8. Sterile PBS: 137 mM NaCl, 2.7 mM KCl, 1.5 μM KH_2PO_4, 8.1 mM Na_2HPO_4 without divalent cations, e.g., $CaCl_2$ or $MgCl_2$.

9. Light microscope with 10× objective.

2.2 Transfection of Endothelial Cells

1. Lipofectamine® 3000 (*see* **Note 5**).

2. Opti-MEM media.

3. Transfection media: Medium 200 supplemented with low serum growth kit (Thermo-Fisher Scientific) and 0.1% FBS (*see* **Note 6**).

4. siRNAs (*see* **Note 7**).

2.3 Dissection and Transfection of Mouse Thoracodorsal Arteries (TDAs)

1. Model organism: mouse.

2. Dissecting microscope with 40× zoom capability.

3. Sterile microdisscetion tools (pair of fine forceps, dissection scissors, etc.)

4. Krebs-HEPES physiological salt solution: 118.4 mM NaCl, 4.7 mM KCl, 1.2 mM $MgSO_4$, 4 mM $NaHCO_3$, 1.2 mM KH_2PO_4, 10 mM HEPES, 6 mM Glucose, 2 mM $CaCl_2$ pH to 7.4 and use with or without 1% Bovine Serum Albumin Fraction V (*see* **Note 8**).

5. Culture myograph 202CM vessel cannulation system (Danish MyoTechnologies) or other vessel cannulation system including glass micropipettes for vessel cannulation.

6. 10-0 Nylon sutures.

7. siRNAs (*see* **Note 7**).

8. Nucleofector™ Cell Line Kit for Human Coronary Artery Endothelial Cells (HCAECs) and Human Aortic Smooth Muscle Cells (hAoSMCs) (Lonza, Swiss).

9. Nucleofector™2b Device (Lonza, Swiss) for cell electroporation or similar device and corresponding electroporation cuvettes.

10. RPMI medium 1640 with or without 1% Bovine Serum Albumin Fraction V, 1% L-glutamine, 1% penicillin/streptomycin (*see* **Note 8**).

11. Incubator at 37 °C, 5% CO_2, 90% humidity, and a class 2 biosafety cabinet (cell culture hood).

12. 100 mm sterile cell culture plastic dishes or 24-well plates.

3 Methods

3.1 Setting Up Plates of Vascular Cells for Experiments

1. For HUVECs only, place 12.5 μL or 25 μL of fibronectin in the middle of each well of a 12- or 6-well plate, respectively, and use a cell scraper to spread the solution around the surface of the well.

2. Leave the plate in the cell culture hood to allow the fibronectin to dry.

3. Once dry, plates can be used immediately or sealed in plastic bags and stored at 4 °C for 1–2 weeks for later use.

4. At 80–90% confluence, remove cell media from flasks of cells and wash cells two times with 5 mL of PBS.

5. Pipette in 1× Trypsin–EDTA. Rock the flask to cover the cells and place back in the 37 °C incubator for 3 min.

6. Check for cell dissociation by microscope.

7. Once 80–90% of cells have become dissociated add back at least twice the amount of media than Trypsin–EDTA to cells. For continued passage add one-fifth of cells back to the flask and top up with fresh media.

8. Count the remaining cells using an hemocytometer and adjust the volume to approximately 2×10^4 cells per mL. This will ensure confluence in 4–5 days (*see* **Note 9**).

9. Place 0.75 or 1.5 mL of cells in each well of a 12- or 6-well plate, respectively, and place plate in 37 °C incubator.

10. Incubate overnight.

3.2 Transfection of HUVEC Cells

1. Remove siRNA from the –80 °C and place on ice to thaw.

2. Prepare 1.5 mL microcentrifuge tubes for siRNA dilution, siRNA, and Lipofectamine additions.

3. Dilute 50 μM stock siRNA 1:10 in Opti-MEM to a concentration of 5 μM (*see* **Note 10**).

4. Prepare siRNA tubes and Lipofectamine tubes by adding 50 μL of room temperature Opti-MEM to each tube per experiment.

5. For 6-well plates (per well), pipette 1 μL of diluted stock siRNA into siRNA tube containing 50 μL Opti-MEM. For 12-well plates (per well), pipette 0.5 μL of diluted stock siRNA into siRNA tube containing 50 μL Opti-MEM.

6. For 6-well plates (per well), pipette 1.0 μL Lipofectamine into the Lipofectamine tube containing 50 μL Opti-MEM. For 12-well plates (per well), pipette 0.5 μL Lipofectamine into the Lipofectamine tube containing 50 μL Opti-MEM.

7. Add the Lipofectamine mix to the siRNA mix and swirl with the pipette tip to mix (*see* **Note 11**).

8. Allow mixed solution to stand for 5–10 min.

9. Remove media from cells and replace with transfection media. For 6-well plates, add 1.9 mL transfection media to each well. For 12-well plates, add 0.9 mL transfection media to each well.

10. Add 100 μL of the siRNA: Lipofectamine mix to each well for a final volume of 2 mL (for 6-well plates) or 1 mL (for 12-well plates) and for a final siRNA concentration of 2.5 nM.

11. Place plate back into the 37 °C incubator for 24 h.

12. After 24 h, remove transfection media and replace with endothelial cell media.

13. Incubate for a further 24 h (*see* **Note 12**).

14. Analyze cell viability and transfection efficiency before using cells in experimental preparations (*see* **Note 13**).

3.3 Dissection of Mouse TDA

1. Euthanize mice (8–10 weeks old) using CO_2 asphyxia and place in the lateral decubitus position at a clean surgical station with dissecting microscope.

2. Spray scapular area with 70 % ethanol to minimize fur getting into the exposed tissue.

3. Make a 3–4 cm long incision on one side in the scapular region using forceps and a dissecting scissor [31].

4. Carefully resect the section of skin covering the latissimus dorsi muscle. Be careful not to disrupt the superficial dorsal muscle underneath the latissimus dorsi (*see* **Note 14**).

5. Remove the latissimus dorsi by microdissection to expose both the superficial dorsal muscle and TDA, which feeds the spino-trapezius muscle. The TDA is surrounded on both sides by veins (*see* **Note 15**).

6. Isolate the TDA (~10–15 mm) by carefully dissecting away surrounding fat and veins. Place TDA in 1.5 mL microcentrifuge tube containing cold Krebs-HEPES on ice. TDA can remain on ice in cold Krebs-HEPES for up to 6 h.

7. Repeat **steps 2–5** for isolating contralateral TDA.

3.4 Ex Vivo Endothelial or Smooth Muscle Cells Transfection

1. Transfer a freshly isolated TDA to the prefilled bath chamber of a pressure myograph (Danish MyoTechnology) with room temperature Krebs-HEPES solution without flow. Make sure pressure myograph tubing and glass micropipettes are prefilled with Krebs-HEPES supplemented with 1 % BSA solution and verify that the tubes and micropipettes do not have any air bubbles (*see* **Notes 8** and **16**).

2. Gently open each end of the TDA with fine forceps and pull each TDA end on opposing glass cannula ends within the chamber. Only handle vessel ends to minimize mechanical stress and float the vessel onto the micropipettes.

3. Secure artery-cannula with 10-0 nylon suture (*see* **Note 17**).

4. Attach the myograph tubing to the pressure transducer and perfuse vessel lumen with Krebs-HEPES 1 % BSA solution for 2 min by turning on the pressure transducer and increasing the inflow pressure to 40 mmHg and set the outflow pressure to 0 mmHg. This will clear residual red blood cells and debris.

5. Mix 82 μL Nucleofection reagent for HCAEC with 18 μL kit supplement 4 and dilute siRNA reagents to final concentration (100 nM or 10 pmol siRNA/sample) in a 1.5 mL microcentrifuge tube (*see* **Note 18**). If only transfecting smooth muscle cells, and not the endothelial cells, use the nucleofection solution for hAoSMC and do not include any siRNA (*see* **Note 19**).

6. Incubate nucleofection solution/siRNA mix for 10 min at room temperature.

7. Disconnect the inflow tubing from the pressure transducer box first and then disconnect that tube from the glass micropipette. This must be performed specifically in this sequence to stop flow through the TDA.

8. Using a 1 mL syringe, completely fill the disconnected inflow tubing with the HCAEC nucleofection solution/siRNA mix. Add a small air bubble to the inflow tube on the side that will be attached to the pressure transducer.

9. Reconnect the inflow tubing to the pressure transducer box and then to the micropipette, specifically in that sequence. Avoid introducing air bubbles directly into the glass micropipette (*see* **Note 20**).

10. Perfuse the vessel lumen with HCAEC nucleofection solution/siRNA by increasing the inflow pressure to 40 mmHg and leaving outflow pressure at 0 mmHg. The bubble near the pressure transducer should move toward the glass micropipette. Allow for the new solution to move through the vessel (the bubble should move at least the length of the glass micropipette to verify the nucleofection solution/siRNA has moved into the vessel (*see* **Note 21**).

11. Quickly and simultaneously remove the vessel from the inflow glass micropipette and tie off the vessel ends with the anchoring suture. Repeat for the opposite end to trap reagents inside vessel lumen.

12. If transfecting only endothelial cells, fill an electroporation cuvette provided with the Nucleofection kit with 100 μL of HCAEC Nucleofection reagent without siRNA (82 μL Nucleofection Reagent for HCAEC with 18 μL kit supplement 4). If transfecting smooth muscle cells, fill the provided electroporation cuvette with hAoSMC Nucleofection reagent (84 μL HAoSMC Nucleofection reagent supplemented with 18 μL of supplement) with siRNA diluted to final concentration. Let this incubate in the cuvette for 10 min at room temperature.

13. Gently transfer the ligated artery to the electroporation cuvette (*see* **Note 22**).

14. Electroporate vessel using program A-034 on Nucleofector 2b device. New protocols for the Nucleofector 4D line are available through Lonza (*see* **Note 23**).

15. Let vessel recover for exactly 5 min in cuvette (*see* **Note 24**).

16. Remove vessel from electroporation cuvette and place in myograph bath filled with Krebs-HEPES solution and carefully remove old sutures with fine forceps. Position vessel ends on opposite micropipettes using new 10-0 nylon sutures. Minimize vessel handling.

17. Flush vessel lumen with Krebes-HEPES 1% BSA solution by increasing the inflow pressure to 40 mmHg for 1 min.

18. Perfuse vessel lumen with warm (37 °C) RPMI 1640 media supplemented with 1% BSA, 1% L-glutamine, and 1% penicillin/streptomycin using an inflow pressure of 40 mmHg by adding this media to the inflow tubing as previously described.

19. Remove vessel from micropipettes (without ligating ends; no sutures) and place in a sterile culture dish or 24-well plate containing RPMI 1640 media supplemented with 1% BSA, 1% L-glutamine, and 1% penicillin/streptomycin (prewarmed to 37 °C) for 18–24 h at 37 °C and 5% CO_2 in humidified cell culture incubator (*see* **Note 25**).

20. Analyze cell viability and transfection efficiency before using vessels in experimental preparations (*see* **Note 13**).

4   Notes

1. To test for off-target effects, nontargeted siRNA are generally used, which consist of a scrambled sequence of the siRNA such that the scrambled sequence doesn't target the target mRNA. This negative control is used to show that the actual procedure including the transfection, introduction of siRNA into the cell, and activation of cellular machinery for RNAi does not affect the mRNA expression or functional output being tested.

2. Primary HUVEC cells are supplied in frozen vials from the manufacturer and should be stored at −80 °C until use.

3. The low serum growth kit from Thermo-Scientific contains 1 mL gentamycin/amphotericin B (500×), a 1 mL mixture of basic fibroblast growth factor (1.5 μg/mL)/heparin (5 mg/mL)/BSA (100 μg/mL), 0.5 mL hydrocortisone (1 mg/mL), 1 mL EGF (5 μg/mL), and 5 mL of FBS. To make the endothelial cell media, we add all of the low serum growth kit with the exception of the FBS (which is not used) to the 500 mL of M200 media, mix for 10 min at room temperature, remove 100 mL (*see* **Note 6**), and replace with 100 mL of FBS that has been batch tested for cell growth.

4. Fibronectin coating of plates promotes adhesion and reduces the tendency of the endothelial cells to float off from the plate surface.

5. Lipofectamine 3000 was selected due to good transfection efficiency as compared to Lipofectamine 2000 in these cells. Other transfection reagents and techniques, e.g., nucleofection may also produce good results in these cells.

6. The 100 mL of no-serum media (removed while making up the stock endothelial cell media) has 50 μL of FBS added to make a final concentration of 0.1 % serum, which is useful for cell transfections and drug treatments and can sustain HUVEC for several days in culture without cell death.

7. Desiccates of siRNA (5 nM) from company should be mixed with 100 μL of DNase free water to a final concentration of 50 μM. siRNAs can then be stored at −80 °C and freeze-thawed up to 50 times without loss.

8. BSA supplemented buffers and media should be sterile filtrated, if they are used in combination with glass micropipettes. Otherwise the BSA will clog the micropipettes.

9. Optimally, cells should never exceed 95 % confluence. Experiments described here require a longer time course, e.g., 3–4 days and should be seeded to reach around 95 % confluence at the end of the experiment. If cells become too confluent, they can detach which will affect the results of the experiments.

10. Recommended usage is 2 nM (1:25,000) to 10 nM (1:5000) from the original 50 μM stock so it is best to dilute the main stock first.

11. Gentle mixing should be used to avoid pipetting up and down.

12. Optimal knockdown occurs between 48 and 72 h after transfection depending on the half-life of the protein of interest (*see* **Note 13**).

13. Cell viability after transfection should be tested and can be assessed by using different commercially available kits. The lactate-dehydrogenase (LDH) Cytotoxicity Assay measures the release of the cytoplasmic enzyme, LDH, into the extracellular space/vessel lumen, which is an indicator for cell death. Trypan blue 0.4 % assay can be used after transfection to monitor the viability of cells. This dye is membrane impermeable to healthy cells but can diffuse across the plasma membrane in dying cells. Additionally, transfection efficiency should be determined. By transfecting fluorophore-coupled non-target siRNAs, the amount of fluorescent cells can be easily determined with a fluorescent microscope. The transfection of a GFP-expressing plasmid can also be used as a control for transfection efficiency, but because of the difference in size, there is no direct comparison possible to the actual amount of siRNA-transfected cells. The successful knockdown of the gene of interest should be determined in at least two ways: on the level of mRNA and protein, for example, via PCR and Western Blot, respectively. The degradation of mRNA could be tested within 24–48 h. It should be taken into account that some ways of gene silencing do not cleave the mRNA and, thus, intact

mRNA could be detected although the translation into protein is inhibited. The detection of the protein of interest via Western Blot is an alternative to test for the mRNA silencing. Depending of the half-life of the investigated protein, protein should be isolated between 48 and 72 h. Connexins with a half-life of 4–5 h could show an effect after 24 h. The appropriate time of knockdown has to be determined for each protein and cell type.

14. Remove enough skin and expose a decent size working area. Keep tissue hydrated by flooding the area with cold Krebs-HEPES supplemented with 1% BSA solution during dissection.

15. TDA is the center vessel of the three parallel blood vessels lying on top of the superficial dorsal muscle.

16. Glass micropipettes should have a diameter around the same size of the TDA vessel or slightly smaller (around 240 μm).

17. Sutures should be preknotted with a single overhand knot and loosely wrapped over the micropipettes before placement of the vessel; two sutures per side are optimal to prevent the vessel from being pulled off of the micropipettes. Tie the sutures over the vessel on the first side that is cannulated and then slide the second side of the vessel onto the other micropipette and tighten the suture over the second side of the vessel.

18. Recommended usage is between 50 and 300 nM (5–30 pmol siRNA/sample).

19. Use HAoSMC kit for smooth muscle cell transfection. In this case, only HAoSMC Nucleofection reagent (with no siRNA) is perfused and trapped in the vessel lumen. For smooth muscle cell transfection, siRNA constructs are mixed in the electroporation cuvette before the vessel is placed inside.

20. When reconnecting the inflow tubing to the glass micropipette, slightly overfill the tube with transfection solution so that a droplet remains on the tip of the tube to be connected to the glass micropipette. Reconnect the inflow tube such that there is no air bubble added to the glass micropipette.

21. Visually observe the filling of the vessel with transfection reagent/siRNA. Only a small amount is necessary to completely fill the vessel.

22. Make sure that the ligated vessel is completely covered with solution.

23. The HCAEC Nucleofection kit is supported by the Nucleofector 2b device and is used for endothelial cell transfection. New reagents and protocols exist for the Nucleofector 4D line of electroporation devices (Lonza) [32].

24. Following transfection, the viability of the vessel is improved when the cells can briefly recover without disturbances for 5 min. However, after 5 min the nucleofection solution may have negative outcomes on cellular viability within the cell.

25. Prewarmed/sterile RPMI media supplemented with 1% BSA, 1% L-glutamine, and 1% penicillin/streptomycin should be equilibrated in a 37 °C incubator for 30 min prior to adding the transfected vessel. In some cases the removal of antibiotics may have favorable outcomes on vessel survival during recovery in cell culture incubator. Recovery times will vary for different types of blood vessels. For TDA, 18 h is a sufficient time point to use vessels for pressure myography experiments. Additionally, the maximal knockdown of certain gene products may occur earlier than 18 h and experimental validation is necessary to show that shorter recovery times do not affect vessel functionality. In this case, measuring an experimental output following the transfection protocol without siRNA or with a negative control scrambled siRNA would be the appropriate control.

Acknowledgements

This work was supported by National Institutes of Health grants (HL088554) and HL120840 and National Institutes of Health training grants (HL007284) (BEI) and by a University of Glasgow, Lord Kelvin Adam Smith Fellowship (SRJ).

References

1. Guo S, Kemphues KJ (1995) par-1, a gene required for establishing polarity in C. elegans embryos, encodes a putative Ser/Thr kinase that is asymmetrically distributed. Cell 81:611–620

2. Fire A, Xu S, Montgomery MK et al (1998) Potent and specific genetic interference by double-stranded RNA in Caenorhabditis elegans. Nature 391:806–811

3. Tabara H, Yigit E, Siomi H et al (2002) The dsRNA binding protein RDE-4 interacts with RDE-1, DCR-1, and a DExH-box helicase to direct RNAi in C. elegans. Cell 109:861–871

4. Chuang CF, Meyerowitz EM (2000) Specific and heritable genetic interference by double-stranded RNA in Arabidopsis thaliana. Proc Natl Acad Sci U S A 97:4985–4990

5. Mourrain P, Beclin C, Elmayan T et al (2000) Arabidopsis SGS2 and SGS3 genes are required for posttranscriptional gene silencing and natural virus resistance. Cell 101:533–542

6. Blevins T, Rajeswaran R, Shivaprasad PV et al (2006) Four plant Dicers mediate viral small RNA biogenesis and DNA virus induced silencing. Nucleic Acids Res 34:6233–6246

7. Choudhary S, Lee HC, Maiti M et al (2007) A double-stranded-RNA response program important for RNA interference efficiency. Mol Cell Biol 27:3995–4005

8. van Rij RP, Saleh MC, Berry B et al (2006) The RNA silencing endonuclease Argonaute 2 mediates specific antiviral immunity in Drosophila melanogaster. Genes Dev 20: 2985–2995

9. Chendrimada TP, Gregory RI, Kumaraswamy E et al (2005) TRBP recruits the Dicer complex to Ago2 for microRNA processing and gene silencing. Nature 436:740–744

10. Wilson RC, Tambe A, Kidwell MA et al (2015) Dicer-TRBP complex formation ensures accurate mammalian microRNA biogenesis. Mol Cell 57:397–407

11. Iwasaki S, Sasaki HM, Sakaguchi Y et al (2015) Defining fundamental steps in the assembly of the Drosophila RNAi enzyme complex. Nature 521:533–536

12. Eamens AL, Smith NA, Curtin SJ et al (2009) The Arabidopsis thaliana double-stranded RNA binding protein DRB1 directs guide strand selection from microRNA duplexes. RNA 15:2219–2235

13. Wilson RC, Doudna JA (2013) Molecular mechanisms of RNA interference. Annu Rev Biophys 42:217–239

14. Doench JG, Petersen CP, Sharp PA (2003) siRNAs can function as miRNAs. Genes Dev 17:438–442

15. Rand TA, Petersen S, Du F et al (2005) Argonaute2 cleaves the anti-guide strand of siRNA during RISC activation. Cell 123:621–629

16. Martinez J, Patkaniowska A, Urlaub H et al (2002) Single-stranded antisense siRNAs guide target RNA cleavage in RNAi. Cell 110:563–574

17. Wei CJ, Francis R, Xu X et al (2005) Connexin43 associated with an N-cadherin-containing multiprotein complex is required for gap junction formation in NIH3T3 cells. J Biol Chem 280:19925–19936

18. Okuda H, Higashi Y, Nishida K et al (2010) Contribution of P2X7 receptors to adenosine uptake by cultured mouse astrocytes. Glia 58:1757–1765

19. Seo M, Lee S, Kim JH et al (2014) RNAi-based functional selection identifies novel cell migration determinants dependent on PI3K and AKT pathways. Nat Commun 5:5217

20. Elbashir SM, Harborth J, Lendeckel W et al (2001) Duplexes of 21-nucleotide RNAs mediate RNA interference in cultured mammalian cells. Nature 411:494–498

21. Takahashi T, Zenno S, Ishibashi O et al (2014) Interactions between the non-seed region of siRNA and RNA-binding RLC/RISC proteins, Ago and TRBP, in mammalian cells. Nucleic Acids Res 42:5256–5269

22. Schwarz DS, Hutvagner G, Du T et al (2003) Asymmetry in the assembly of the RNAi enzyme complex. Cell 115:199–208

23. Parsons BD, Schindler A, Evans DH et al (2009) A direct phenotypic comparison of siRNA pools and multiple individual duplexes in a functional assay. PLoS One 4:e8471

24. Lohman A, Straub A, Isakson B (2012) Endothelial cell transfection of ex vivo arteries. Protocol Exchange. doi:10.1038/protex.2012.1052

25. Begandt D, Bader A, Antonopoulos GC et al (2015) Gold nanoparticle-mediated (GNOME) laser perforation: a new method for a high-throughput analysis of gap junction intercellular coupling. J Bioenerg Biomembr 47:441–449

26. Heinemann D, Schomaker M, Kalies S et al (2013) Gold nanoparticle mediated laser transfection for efficient siRNA mediated gene knock down. PLoS One 8:e58604

27. Zentilin L, Giacca M (2004) In vivo transfer and expression of genes coding for short interfering RNAs. Curr Pharm Biotechnol 5:341–347

28. Sliva K, Schnierle BS (2010) Selective gene silencing by viral delivery of short hairpin RNA. Virol J 7:248

29. Yuan D, Sun G, Zhang R et al (2015) Connexin 43 expressed in endothelial cells modulates monocyteendothelial adhesion by regulating cell adhesion proteins. Mol Med Rep 12:7146–7152

30. O'Donnell JJ 3rd, Birukova AA, Beyer EC et al (2014) Gap junction protein connexin43 exacerbates lung vascular permeability. PLoS One 9:e100931

31. Billaud M, Lohman AW, Straub AC et al (2012) Characterization of the thoracodorsal artery: morphology and reactivity. Microcirculation 19:360–372

32. Ketsawatsomkron P, Lorca RA, Keen HL et al (2012) PPARgamma regulates resistance vessel tone through a mechanism involving RGS5-mediated control of protein kinase C and BKCa channel activity. Circ Res 111:1446–1458

Chapter 6

Generation and Use of Trophoblast Stem Cells and Uterine Myocytes to Study the Role of Connexins for Pregnancy and Labor

Mark Kibschull, Stephen J. Lye, and Oksana Shynlova

Abstract

Transgenic mouse models have demonstrated critical roles for gap junctions in establishing a successful pregnancy. To study the cellular and molecular mechanisms, the use of cell culture systems is essential to discriminate between the effects of different connexin isoforms expressed in individual cells or tissues of the developing conceptus or in maternal reproductive tissues. The generation and analysis of gene-deficient trophoblast stem cell lines from mice clearly revealed the functions of connexins in regulating placental development. This chapter focuses on the use of connexin gene-deficient trophoblast stem cell cultures to reveal the individual role of gap junctions in regulating trophoblast differentiation and proliferation in vitro under controlled conditions. In addition, cultures of primary uterine myocytes, isolated from mice or rats, allow studying the effects of mechanical stretch or ovarian hormones on regulating connexin expression, and thus, to model the molecular mechanisms of uterine growth and development during pregnancy. Here, we describe the derivation of primary uterine myocyte cultures and their use in in vitro stretch experiments to study the mechanisms of myometrial remodeling essential to accommodate the growing fetus throughout gestation.

Key words Placenta, Trophoblast differentiation, Embryo, Uterus, Myometrium, Gestation, Stretch

1 Introduction

During pregnancy, the developing embryo establishes complex interactions with maternal uterine tissues. Physiological, genetic, or environmental impairment of the fetomaternal interaction can lead to pregnancy complications, such as preeclampsia, intrauterine growth restriction, or preterm labor, affecting the health of both the baby and the mother [1, 2]. Studying developmental and molecular mechanisms of the placenta or the pregnant uterus is challenging due to the presence of tissues of two genetically divergent organisms. Current experimental approaches include the use of different animal models to examine specific cellular, hormonal, or immunological aspects of fetal and/or maternal development.

Mathieu Vinken and Scott R. Johnstone (eds.), *Gap Junction Protocols*, Methods in Molecular Biology, vol. 1437,
DOI 10.1007/978-1-4939-3664-9_6, © Springer Science+Business Media New York 2016

Mutant mice have demonstrated that gap junctions play essential roles in establishing a successful pregnancy from regulating trophoblast differentiation during placental development and mediating uterine artery remodeling [3, 4], and to forming a functional myometrial syncytium, which enables forceful uterine contractions to expel the fetus during labor [5]. However, due to the presence of multiple connexin isoforms expressed by the placental trophoblasts (i.e., Cx31, C31.1, Cx26, and Cx43), endothelial cells (i.e., Cx40, Cx43, and Cx37), the decidua (i.e., Cx43), the fetal mesoderm (i.e., Cx43 and Cx45), and by uterine myocytes (i.e., Cx26, Cx43, and Cx45) [3], the specific effects of individual connexins are difficult to study. The derivation of tissue cell lines and long-term or short-term primary cell cultures that mimic the physiological processes in utero allow studying the molecular mechanisms leading to a successful gestation.

In the developing placenta, Cx31 and Cx31.1 are critical regulators of trophoblast differentiation. Both connexins are coexpressed in the proliferating postimplantation trophoblast lineage. The inactivation of either gene in mice results in partial loss (i.e., 60 and 30%, respectively) of conceptuses between embryonic days 10.5 and 13.5 [6, 7]. Surviving embryos are born viable, but growth restricted due to significantly reduced placental sizes [7, 8]. For both mutant mouse strains, the generation of Cx31 or Cx31.1 gene-deficient trophoblast stem cell lines and their in vitro analysis has provided fundamental insights into the role of these gap junction proteins in regulating trophoblast differentiation (Fig. 1) [9, 10].

Trophoblast stem cell (TSC) lines can be generated from a blastocyst's trophectoderm outgrowth or dissected extraembryonic ectoderm in the presence of fibroblast growth factor 4 (FGF4) and mouse embryonic fibroblast conditioned medium (MEF-CM). A later study showed that transforming growth factor beta/activin is the critical factor produced by MEFs [11]. By removing FGF4 and MEF-CM/activin from cell culture medium, TSC can differentiate into all placental trophoblast subpopulations [12, 13]. Earlier, we showed that during in vitro differentiation, wild-type TSC maintain a specific placental connexin expression pattern. In particular, undifferentiated TSC express both Cx31 and Cx31.1, and upon differentiation into syncytiotrophoblast, the expression of Cx26 is induced, while differentiation into trophoblast giant cells stimulates the expression of Cx43 [10, 14]. The generation of Cx31-deficient and Cx31.1-deficient TSC lines from blastocysts of the corresponding

Fig. 1 (continued) Cx31 preserves the diploid, proliferating trophoblasts, whereas Cx31.1 promotes differentiation. (**b–e**) *Quantitative real-time polymerase chain reaction analysis of each 3 Cx31.1+/− (HZ) and Cx31.1−/− (KO) TSC lines differentiated* in vitro *for 6 days.* Cx31.1 deficiency results in a block of terminal differentiation from the Mash2-expressing intermediate trophoblast population into placental glycogen cells, Pcdh12, spongiotrophoblast, Tpbpa, and giant cells, PI-2 as indicated by repressed marker gene expression compared to Cx31.1+/− (HZ) controls

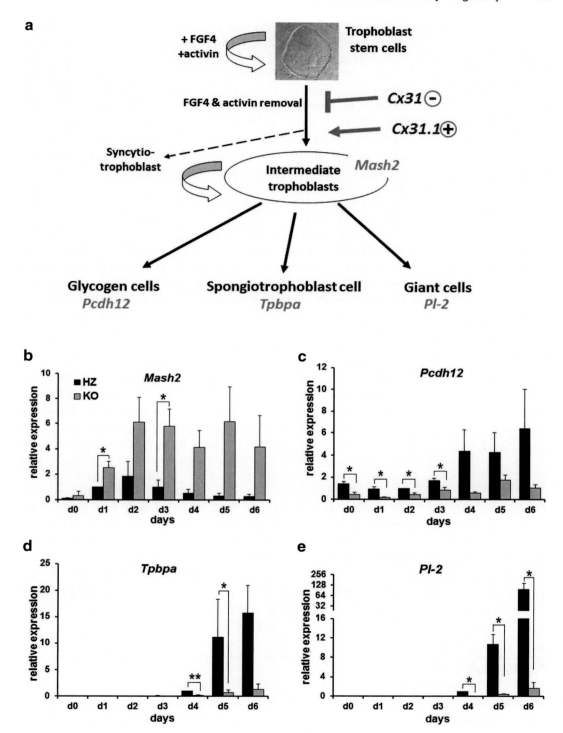

Fig. 1 (**a**) *Scheme demonstrating the opposite roles of Cx31 and Cx31.1 in regulation TSC differentiation.* Upon removal of FGF4 and activin, TSC differentiate into intermediate trophoblasts characterized by Mash2 expression and further into the placental trophoblast subpopulation, namely, glycogen cells (protocadherin12, Pcdh12), spongiotrophoblast (trophoblast binding protein A, Tpbpa), and trophoblast giant cells (placental lactogen 2, Pl-2).

knock-out (KO) mice revealed an antipodal role for both connexin isoforms. Cx31-deficient TSC differentiate faster along the trophoblast lineage compared to wild-type control cells [10], whereas Cx31.1-deficient TSC differentiate slower into placental trophoblast subpopulations compared to controls [9], as indicated by the analysis of specific marker genes and proliferation capabilities (Fig. 1). This clearly indicates that Cx31 is critical to maintaining TSCs in the undifferentiated proliferative state during placental development and the coexpression of Cx31.1 has an opposite function of promoting trophoblast lineage differentiation into placental subpopulations [9, 10]. Therefore, TSC cultures provide a unique and valuable tool for investigating the role of gap junction proteins in lineage differentiation during placental development.

During pregnancy, the uterus undergoes dramatic changes to accommodate the growing fetus. The uterine smooth muscle (i.e., myometrium) is noticeably increasing in size. This growth is regulated by mechanical stretch of the uterine wall by the growing fetus(es) and by pregnancy-related changes of the ovarian hormones estrogen and progesterone [15]. In addition, uterine remodeling is critically regulated by the invading trophoblast populations, paracrine and endocrine signals, and by the infiltration of immune cells into the uterine tissues [16, 17]. The expression of gap junction proteins (i.e., Cx43, Cx26, and Cx45) is regulated throughout gestation to ensure tissue integrity during the development of cellular hypertrophy and later for coordinated forceful uterine contractions during labor to expel the fetus/es [18]. Tissue-specific deletion of Cx43 from uterine myocytes in mice leads to irregular uterine contractions and delayed delivery of pups, revealing the importance of gap junctional communication for successful pregnancies [19].

The establishment of primary rodent myocyte cultures allows to analyze in vitro the effect of mechanical stretch and/or hormonal changes on gap junctional communication between uterine myocytes independent of vascular cells, resident or infiltrating leukocytes or fetal tissues [19, 20].

This chapter describes the development of different cell culture systems derived from fetal (i.e., trophoblasts) and maternal (i.e., myocytes) rodent tissues to study the complex physiological mechanisms underlying pregnancy.

2 Materials

2.1 Culture and Differentiation of Mouse Trophoblast Stem Cells

1. TS medium: Roswell Park Memorial Institute (RPMI) medium 1640, 20 % fetal bovine serum (FBS) (Thermo-Fisher Scientific, USA) (see Note 1), 2 mM glutamine, 1 mM sodium pyruvate, 100 mM β-mercaptoethanol, and 100 U/mL penicillin/streptomycin.

2. Mouse embryonic fibroblast (MEF) conditioned TS medium (TS/CM) [14] (*see* **Note 2**).

3. TSC lines cultured in 70% MEF conditioned medium (TS/CM) and 30% TS medium, supplemented with 25 ng/mL FGF4 (R&D Systems, USA) (*see* **Note 3**) and 1 U/mL heparin (Sigma-Aldrich, Canada).

4. TrypLE dissociation reagent (Life Technologies, Canada) (*see* **Note 4**).

5. Incubator at 37 °C, 5% CO_2, 5% CO_2 and 90% humidity, and biosafety cabinet.

6. Cell culture plastic dishes and 40 μm cell strainer.

7. Dulbecco's phosphate buffered saline (DPBS) at pH 7.4 (Sigma-Aldrich, Canada).

2.2   Colorimetric TSC Proliferation Assay

1. Staining solution: 1.6 mg/mL 3-(4,5-dimethylthiazol-2-yl)-2,5-diphenyltetrazolium bromide (MTT) (Sigma-Aldrich, Canada) dissolved in DPBS.

2. 24-Well culture plates.

3. Dimethylsulfoxide (DMSO).

4. Plate reader photometer.

2.3   Derivation and Culture of Primary Rodent Uterine Myocytes

1. Estrogen solution: 0.25 mg/mL 17β-estradiol (Sigma-Aldrich, Canada) dissolved in 90% corn oil with 10% ethanol.

2. Buffer A: sterile Hank's Basic Salt Solution (HBSS) containing calcium and magnesium, 25 mM 2-[4-(2-hydroxyethyl)piperazin-1-yl]ethanesulfonic acid (HEPES), 100 U/mL penicillin/streptomycin, and 2.5 μg/mL amphotericin B. Adjust to pH 7.4.

3. Buffer B: buffer A without calcium and magnesium.

4. Buffer C: buffer B supplemented with 1 mg/mL collagenase type II (Sigma-Aldrich, Canada), 0.15 mg/mL DNase I (Roche, Canada), 0.1 mg/mL soybean trypsin inhibitor, 10% FBS (Cansera, Canada), and 1 mg/mL bovine serum albumin (BSA).

5. Cell culture medium: phenol red-free Dulbecco's Modified Eagle's Medium (DMEM) (Life Technologies, Canada) supplemented with 10% FBS, 25 mM HEPES, 100 U/mL penicillin/streptomycin, and 2.5 μg/mL amphotericin B.

6. Serum-free medium: phenol red-free DMEM supplemented with insulin-, transferrin-, selenium-, sodium pyruvate solution (ITS-A) (Life Technologies, Canada), 25 mM HEPES, 100 U/mL penicillin/streptomycin, and 2.5 μg/mL amphotericin B.

2.4   In Vitro Stretch of Primary Uterine Myocyte Cultures

1. Flexcell FX-3000 Tension System (Flexcell Inc., USA).

2. Collagen I-coated Bio-Flex 6-well plates with rubber membranes (Flexcell Inc., USA).

3. ProNectin-coated 6-well culture plates with rubber membranes (Flexcell Inc., USA).

4. Collagen I-coated HT Bio-Flex 24-well plates with rubber membranes (Flexcell Inc., USA).

5. ProNectin-coated HT Bio-Flex 24-well plates with rubber membranes (Flexcell Inc., USA).

6. Covalently bound matrix surfaces: amino, collagen (i.e., type I or IV), elastin and pronectin (Flexcell Inc., USA).

3   Methods

3.1 Comparative Analysis of Connexin-Deficient TSC Differentiation

1. Use gene-deficient TSC lines and wild-type control lines, which were derived in parallel (*see* **Note 5**) using the same batch of TS/CM (*see* **Note 3**). Change cell culture media every other day.

2. At 80% confluency, remove cell culture medium, add 1 mL TrypLE, and incubate for 3 min at room temperature. Dissociate TSC to single cells by pipetting up and down with a 1 mL pipette tip. Add 1 mL of TS medium and pass the suspension through a 40 μm cell strainer to remove clumps and aggregates of differentiated cells.

3. Spin the suspension at $800 \times g$ for 3 min, remove supernatant, and resuspend cells in 1 mL of TS/CM. Count cells and seed in 12-well plates at a density of 10,000 cells *per* well in 1 mL of TS/CM. Incubate plates for 48 h before inducing differentiation (*see* **Note 7**).

4. To induce TSC differentiation, remove the cell culture medium and wash monolayers with DPBS to remove any traces of TSC/CM. Add 1 mL of TS medium and culture plates for up to 10 days (*see* **Note 6**).

5. Take samples at days 0, 2, 4, 6, 8, and 10. Isolate RNA and perform standard quantitative real-time polymerase chain reaction analysis for specific trophoblast differentiation markers (Fig. 1).

3.2 MTT-Proliferation Assay for TSC

1. Plate TSC in 24-well plastic plates at a density of 10,000 cells *per* well in 0.5 mL of TSC/CM. Use four technical replicates for each measuring point. Culture plates for 48 h in the incubator.

2. At 48 h postplating, establish the reference point for the proliferation curve. Remove cell culture medium and add 1 mL of prewarmed 10% staining solution in TS medium to the wells at a final MTT concentration 160 μg/mL (*see* **Note 7**). Incubate plates for 2 h in the incubator.

3. Carefully remove medium and add 0.5 mL of DMSO to each well. Shake the plates for 5 min at low speed on a shaker to lyse cells and release the blue dye.

4. Place the plates into a plate reader or transfer solutions into cuvettes and measure in a photometer at 570 nm wave length with DMSO as a background (i.e., blank).

5. Measure four technical replicates for each time point (i.e., day) of the proliferation curve.

6. Normalize data to day 0, graph and analyze using appropriate computer software (Fig. 2).

3.3 Derivation and Culture of Primary Rodent Uterine Myocytes

Primary rodent uterine myocytes are prepared as follows [21, 22]:

1. Virgin, female, Sprague-Dawley rats weighing 150–200 g are subcutaneously injected with estrogen (i.e., 50 μg in 200 μL).

2. 24 h after injection, excise the whole uterus under sterile conditions and place in a 10 cm cell culture dish with 25 mL of buffer A.

3. Place the sterile cell culture dish with uterine tissue in a biosafety cabinet with laminar flow, clean the uterine horns from fat and connective tissue, cut into 1 mm wide rings, and place them in a 50 mL flask with 25 mL of buffer B.

4. Wash the tissue pieces three times with buffer B at room temperature.

5. Perform an enzymatic digestion of the tissue by incubation in buffer C (i.e., 10 mL/g of tissue) for 30 min at 37 °C on an orbital shaker (i.e., 100 rounds *per* min).

6. Following incubation, gently agitate the mixture by repeated trituration with a glass and plastic pipette to mechanically disrupt uterine tissue (*see* **Note 8**).

7. Add an equal volume of ice-cold buffer B with 10 % FBS to stop enzymatic digestion, pass the suspension through a 70 μm cell strainer to remove clumps of cells, and store on ice.

8. Put the remaining undigested tissue into a new flask, add 10 mL of fresh buffer C to the tissue, and repeat the incubation and aspiration process five times.

9. Collect the dissociated cells from **steps 2–6** by centrifugation at $200 \times g$ for 15 min and resuspend the cell pellet in DMEM supplemented with 10 % FBS (*see* **Note 9**).

3.4 Application of Stretch to Rat Smooth Muscle Cells

1. Plate freshly isolated myometrial cells with an initial density of 3×10^6 cells *per* well in 3 mL of cell culture medium into 6-well Flexcell plates coated with type I collagen or other extracellular matrix proteins. Leave for 3 days to attach and proliferate in a humidified 5 % CO_2 incubator at 37 °C to 75 % confluence.

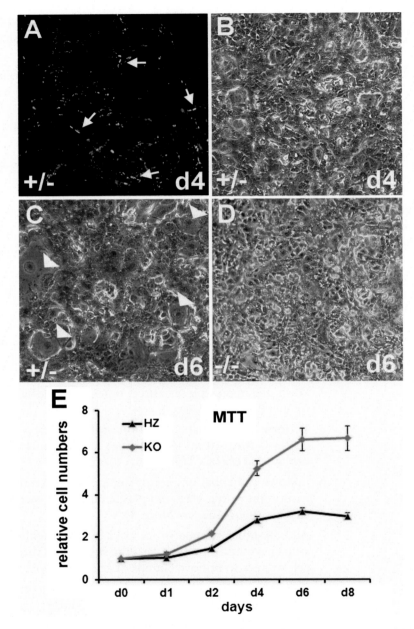

Fig. 2 (**a**) Immunolabeling for Cx31.1 (*arrows*) and (**b**) corresponding phase-contrast image of Cx31.1$^{+/-}$ TSC at day 4 of differentiation. (**c**) At day 6, giant cells (*arrow heads*) are prominent in Cx31.1$^{+/-}$ TSC cultures, whereas the delayed differentiation of Cx31.1$^{+/-}$ TSC (**d**) leads to absence of giant cells forming cultures at day 6. Blue LacZ stain indicates the Cx31.1 expression cell populations. (**e**) MTT proliferation assay shows increased proliferation rates of Cx31.1$^{-/-}$ during differentiation compared to Cx31.1$^{+/-}$ controls (magnification 10×)

2. Incubate the cells for 24 h in serum-free medium to render quiescence.

3. Expose the cell culture plates to static stretch ranging between 0 and 25 % elongation for predetermined time points (i.e., 2,

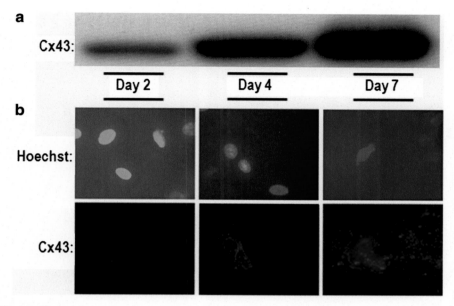

Fig. 3 *Basal Cx43 protein levels in myometrial smooth muscle cells following 2, 4, and 7 days in culture.* (**a**) Representative immunoblot analysis showing the dramatic increase in Cx43 protein levels with increasing days in culture. (**b**) Indirect immunofluorescence of Cx43 in myometrial smooth muscle cells showing intense punctate staining for Cx43 on day 7 of culture. Cells were simultaneously stained with Hoechst to mark cell nuclei (magnification 1000×)

6, or 24 h) by applying a vacuum generated by a pump and controlled by a computer-driven system (*see* **Note 10**).

4. Extract proteins or RNA, or fix with 4 % paraformaldehyde (*see* **Note 11**).

5. Analyze gap junction expression (Fig. 3) and viability of cells (Fig. 4).

4   Notes

1. The source of FBS and percentage of conditioned TS medium have major effects on the undifferentiated growth of TSC. FBS should be tested to promote undifferentiated growth of TSC over several passages. Depending on the performance of FBS, the concentration of CM might be reduced [14].

2. We routinely use MEFs derived from the same mouse strain as the TSC lines. The autologous MEFs always lead to successful generation and culture of TSC lines.

3. It is crucial for comparison of differentiation and proliferation of mutant and wild-type TSC to use the same batch of TS medium as well as 70 % TS/CM. We strongly recommend preparing large volumes of TS/CM and freezing at −20 °C. The

Control (non-stretched)   Stretched for 24 hours

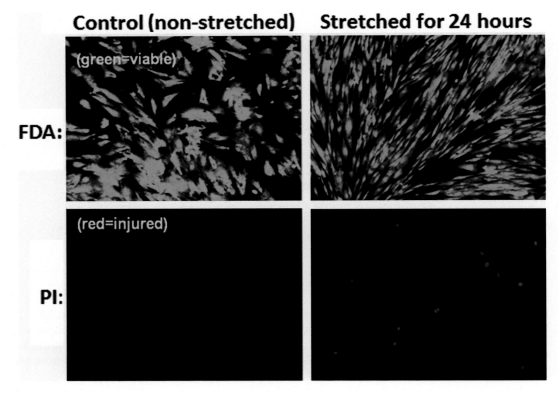

FDA: (green=viable)

PI: (red=injured)

Fig. 4 *Static mechanical stretch applied by Flexcell FX-3000 does not induce cell injury.* Cell viability assay by fluorescein diacetate (FDA)-propidium iodide (PI) staining of primary rat myometrial smooth muscle cells. FDA (*green*) and PI (*red*) staining of nonstretched cells (control) and primary uterine myocytes following 24 h of static stretch (25 % elongation) (magnification 200×)

quality of FGF4 has major effects on undifferentiated growth of TSC. Vendors should provide quality testing of FGF4 activity on data sheets, which is usually a sign of quality.

4. TrypLE is a recombinant cell dissociation enzyme, which is more effective on TSC compared to porcine trypsin-based solutions. In particular, when using TrypLE, dissociation of TSC to a single cell suspension is much faster and more reproducible.

5. For the differentiation assay, using connexin-deficient TSC is important to compare mutant and control TSC derived from the same mouse strain. Ideally, TSC should be derived in 1 experiment using the same batch of feeders and TS media. This approach will help to identify true differences in the marker gene profile during cell differentiation. TSC derived from different mouse strains (i.e. SV129 and C57BL/J) show significant difference in expression levels of marker genes, though the temporal expression profile is still comparable. We routinely mate heterozygous females with homozygous males for

blastocyst isolation to simultaneously generate several heterozygous and homozygous connexin-deficient TSC lines. When TSC from different rounds of derivation are used, we recommend first to normalize the data of mutant TSC to corresponding controls before combining all data into a study.

6. We do not recommend plating TSC in TS medium directly to induce differentiation. Seeding TSC in TSC/CM and allowing them to adhere and form initial colonies will significantly increase cell attachment and reduce spontaneous differentiation. Thus, for better performance and less variability of differentiation results as well as proliferation assays, we strongly recommend 48 h of incubation time.

7. MTT at high concentrations (i.e., 0.5 mg/mL) normally used for cancer cell lines is stressful to TSC and leads to detachment of cells over time. We found that an MTT concentration of 60 µg/mL and using a 2 h incubation time led to reproducible experimental results of TSC proliferation assays without losing cells. The MTT assay is most useful when studying proliferation during TSC differentiation, as differentiated trophoblast cells are hard to detach enzymatically and therefore cannot be reliably counted as dissociated cells.

8. We recommend at the end of the incubation cycle (i.e., 30 min) that the mixture is gently agitated by repeated titration (i.e., 3–4 min) with a 25 mL large-hole glass pipette to aid enzymatic dispersion of uterine tissue. Due to the decreasing size of tissues during the procedure, large-hole and small-hole plastic transfer pipettes can be used for the final two incubation steps, respectively.

9. The first incubation solution is discarded, since it contains debris and damaged cells. To selectively enrich for uterine myocytes, we recommend subjecting the freshly isolated cell suspension to a differential attachment technique. For this purpose, preplate dissociated cells on polystyrene culture dishes for 30–45 min at 37 °C, during which period the quickly adhering nonmyocytes, mostly fibroblasts, will readily attach to the bottom of the cell culture dish. The supernatant containing slowly adhering uterine smooth muscle cells should be collected and plated on the Flexcell plates. Both cell count and cell viability could be assessed by trypan blue exclusion using a hemocytometer.

10. The Flexcell strain unit has been characterized in detail [23]. It consists of a vacuum unit regulated by a solenoid valve and a computer program. When a precise vacuum level is applied to the system, the cell culture plate bottoms are deformed in downward direction to a known percentage elongation, which is translated to the cultured cells. When the vacuum is released,

the plate bottoms return to their original conformation. The magnitude, duration, and frequency of the applied force can be varied in this system. The force on the attached cells is predominantly uniaxial. However, the deformation of the flexible membrane is not uniform, but rather generates a gradient stretch, with the greatest deformation occurring at the periphery. Therefore, the results of the stretch experiment represent an average of cells exposed to different degrees of stretch. For 25 % elongation, the average elongation is approximately 10 % over the entire cell culture plate surface.

11. Stretched and control (i.e., nonstretched) Flexcell plates should be established simultaneously with the same pool of cells in each experiment to match for temperature, CO_2 content, or pH of the cell culture medium.

Acknowledgements

We gratefully thank Mrs. Alexandra Oldenhof for assistance in collecting and processing of rat tissues. This work was supported by a grant from the Deutsche Forschungsgesellschaft (DFG, KI1278/1-1) to M.K., and from the Canadian Institute of Health Research (CIHR, MOP-37775) to S.J.L. and O.S.

References

1. Roberts DJ, Post MD (2008) The placenta in pre-eclampsia and intrauterine growth restriction. J Clin Pathol 61:1254–1260

2. Romero R, Dey SK, Fisher SJ (2014) Preterm labor: one syndrome, many causes. Science 345:760–765

3. Kibschull M, Gellhaus A, Carette D et al (2015) Physiological roles of connexins and pannexins in reproductive organs. Cell Mol Life Sci 72:2879–2898

4. Winterhager E, Kidder GM (2015) Gap junction connexins in female reproductive organs: implications for women's reproductive health. Hum Reprod Update 21:340–352

5. Shynlova O, Lee YH, Srikhajon K et al (2013) Physiologic uterine inflammation and labor onset: integration of endocrine and mechanical signals. Reprod Sci 20:154–167

6. Plum A, Winterhager E, Pesch J et al (2001) Connexin31-deficiency in mice causes transient placental dysmorphogenesis but does not impair hearing and skin differentiation. Dev Biol 231:334–347

7. Zheng-Fischhofer Q, Kibschull M, Schnichels M et al (2007) Characterization of connexin31.1-deficient mice reveals impaired placental development. Dev Biol 312:258–271

8. Kibschull M, Magin TM, Traub O et al (2005) Cx31 and Cx43 double-deficient mice reveal independent functions in murine placental and skin development. Dev Dyn 233:853–863

9. Kibschull M, Colaco K, Matysiak-Zablocki E et al (2014) Connexin31.1 (Gjb5) deficiency blocks trophoblast stem cell differentiation and delays placental development. Stem Cells Dev 23:2649–2660

10. Kibschull M, Nassiry M, Dunk C et al (2004) Connexin31-deficient trophoblast stem cells: a model to analyze the role of gap junction communication in mouse placental development. Dev Biol 273:63–75

11. Erlebacher A, Price KA, Glimcher LH (2004) Maintenance of mouse trophoblast stem cell proliferation by TGF-beta/activin. Dev Biol 275:158–169

12. Quinn J, Kunath T, Rossant J (2006) Mouse trophoblast stem cells. Methods Mol Med 121:125–148

13. Tanaka S, Kunath T, Hadjantonakis AK et al (1998) Promotion of trophoblast stem cell proliferation by FGF4. Science 282:2072–2075

14. Kibschull M, Winterhager E (2006) Connexins and trophoblast cell lineage development. Methods Mol Med 121:149–158

15. Shynlova O, Tsui P, Jaffer S et al (2009) Integration of endocrine and mechanical signals in the regulation of myometrial functions during pregnancy and labour. Eur J Obstet Gynecol Reprod Biol 144:S2–S10

16. Dunk CE, Gellhaus A, Drewlo S et al (2012) The molecular role of connexin 43 in human trophoblast cell fusion. Biol Reprod 86:115

17. Shynlova O, Nedd-Roderique T, Li Y et al (2013) Infiltration of myeloid cells into decidua is a critical early event in the labour cascade and post-partum uterine remodelling. J Cell Mol Med 17:311–324

18. Orsino A, Taylor CV, Lye SJ (1996) Connexin-26 and connexin-43 are differentially expressed and regulated in the rat myometrium throughout late pregnancy and with the onset of labor. Endocrinology 137:1545–1553

19. Doring B, Shynlova O, Tsui P et al (2006) Ablation of connexin43 in uterine smooth muscle cells of the mouse causes delayed parturition. J Cell Sci 119:1715–1722

20. Shynlova O, Nedd-Roderique T, Li Y et al (2013) Myometrial immune cells contribute to term parturition, preterm labour and post-partum involution in mice. J Cell Mol Med 17:90–102

21. Shynlova OP, Oldenhof AD, Liu M et al (2002) Regulation of c-fos expression by static stretch in rat myometrial smooth muscle cells. Am J Obstet Gynecol 186:1358–1365

22. Mollard P, Mironneau J, Amedee T et al (1986) Electrophysiological characterization of single pregnant rat myometrial cells in short-term primary culture. Am J Physiol 250:C47–C54

23. Gilbert JA, Weinhold PS, Banes AJ et al (1994) Strain profiles for circular cell culture plates containing flexible surfaces employed to mechanically deform cells in vitro. J Biomech 27:1169–1177

Chapter 7

Identification of Connexin43 Phosphorylation and *S*-Nitrosylation in Cultured Primary Vascular Cells

Alexander W. Lohman, Adam C. Straub, and Scott R. Johnstone

Abstract

All connexins (Cx) proteins contain both highly ordered domains (i.e., 4 transmembrane domains) and primarily unstructured regions (i.e., n- and c-terminal domains). The c-terminal domains vary in length and amino acid composition from the shortest on Cx26 to the longest on Cx43. With the exception of Cx26, the c-terminal domains contain multiple sites for posttranslational modification (PTM) including serines (S), threonines (T), and tyrosines (Y) for phosphorylation or cysteines (C) for *S*-nitrosylation. These PTMs are critical for regulating cellular localization, protein–protein interactions, and channel functionality. There are several biochemical techniques that allow for the identification of these PTM including Western blotting and the "Biotin Switch" assay for nitrosylation. Quantitative analysis of Western blots can be achieved through use of secondary antibodies in the near infrared scale and high-resolution scanning on a fluorescent scanner.

Key words Biotin switch, Fluorescence-based Western blotting, Membrane preparation, Phosphorylation, Posttranslational modification, *S*-nitrosylation

1 Introduction

Blood vessels are composed of two main layers of cells, the endothelial cells (EC) which line the inner or luminal surface and the smooth muscle cells (SMC) surrounding the ECs, separated by layers of matrix proteins. The endothelial layer plays an important role in maintaining vessel integrity and sensing the surrounding environment through contact with circulating blood cells. The smooth muscle layers are integral in vessel structure and contractility. The composition of the layers of the blood vessels varies highly along the vascular tree. In smaller resistance arteries, a thinner layer of matrix proteins allows for heterocellular communication between EC and SMC through a domain called the myoendothelial junction which is not present in large conduit arteries [1]. In both large and small vessels, gap junctions play key roles in conducting signals between EC and SMC and along the length of the vessel.

Mathieu Vinken and Scott R. Johnstone (eds.), *Gap Junction Protocols*, Methods in Molecular Biology, vol. 1437,
DOI 10.1007/978-1-4939-3664-9_7, © Springer Science+Business Media New York 2016

In the vasculature there are five identified connexin (Cx) iso-forms, Cx32, Cx37, Cx40, Cx43, and Cx45, which regulate the coordination of vessel contraction and relaxation. In general terms, Cx32, Cx37, and Cx40 are the most abundant in ECs with Cx43 and Cx45 routinely identified in the vascular SMCs [1]. However, this expression pattern varies depending on vessel size and function with Cx43 found in the endothelium at arterial branch points and in smaller resistance vessels [2].

In order to control vascular functions, gap junction communi-cation must be regulated at multiple levels including trafficking of connexin hemichannels (the building blocks for gap junctions) to the plasma membrane, docking of opposed hemichannels to form patent gap junctions, and modulation of channel permeability. While a number of factors including intracellular pH and calcium concentrations can regulate these events, there is substantial evi-dence that membrane trafficking, gap junction aggregation, pro-tein interactions, and channel gating are regulated by PTMs such as phosphorylation and nitrosylation [3–6]. In this chapter, we will focus on PTMs of Cx43, although there is clear evidence that other vascular connexins (e.g. Cx37 and Cx40) undergo similar modifi-cations in the vasculature [7].

Phosphorylation encompasses the addition of phosphate groups to polar amino acid side chains, primarily serine (S), threonine (T), and tyrosine (Y) residues. For Cx proteins phosphorylation is facili-tated through the action of multiple protein kinases with differential affinity for these residues [5, 8]. For example, protein kinase C (PKC), cAMP-dependent protein kinase (PKA), mitogen-activated protein kinases (MAPK), cyclin-dependent kinases (Cdk), casein kinase (CK1), and calmodulin-dependant protein kinases (CaMK) preferentially phosphorylate serine/threonine residues whereas Src family kinases (SFK) preferentially target tyrosine residues [9]. The intracellular loop of Cx43 does not contain sites for phosphorylation and is not thought to be modified; however, the amino terminus has a single serine (S5) which could potentially be modified. An example of the phosphorylated regions and associated kinases for Cx43 c-terminus is shown in Fig. 1 [10]. Western blotting for Cx43 typi-cally reveals multiple bands ranging in molecular weight between 37kD and 50kD. These bands represent differentially phosphory-lated species and are most commonly denoted in the literature as the NP/P0 (null-phosphorylated), P1 and P2 (phosphorylated) bands (Fig. 2). However, this nomenclature is not strictly true, as phos-pho-specific antibodies to pS368 a PKC site, colocalizes with the P0 band [11, 12]. Multiple factors appear to alter banding pattern including cell type, cell confluence, protein lysate preparation, pri-mary antibodies used and differences in detection methods. In addi-tion higher and lower migrating bands are commonly seen but have never been fully explained (Fig. 3) [6]. Phosphorylation is thought to be important for Cx hemichannel incorporation into gap

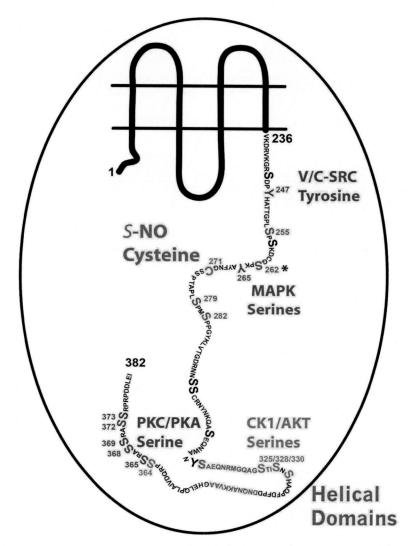

Fig. 1 *Schematic of the Cx43 protein in the plasma membrane.* Experimentally identified sites of phosphorylation and nitrosylation are highlighted. Sites in bold and black font are based on prediction software and have not been shown experimentally. * The Serine 262 site has also been shown to be phosphorylated through PKC-ε. Image is an adaption from Campbell et al. [10]

junctions and function of Cx43 in the cell membrane. Phosphorylated connexin proteins are found in cell membranes and specific cellular compartments, thus it can be useful to isolate proteins from membrane fractions, large organelles, and cytosolic proteins for comparisons (Fig. 4). In order to separate out these fractions, a membrane preparation adapted from Berg et al. can be employed [13].

Nitric oxide (NO) is a potent vasodilator and can alter protein functions through another form of PTM called *S*-nitrosylation. Similar to phosphorylation, *S*-nitrosylation acts to covalently

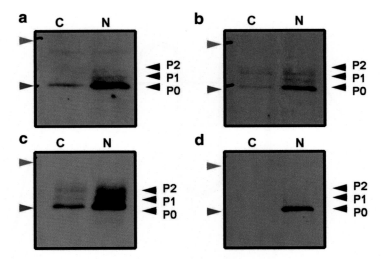

Fig. 2 *Identification of Cx43 phophorylated bands using different antibodies.* Antibodies directed against c-terminal regions for Cx43 were purchased from Sigma monoclonal Cx43 (**a**), Sigma polyclonal Cx43 (**b**), Zymed polyclonal (**c**), and ADI polyclonal (**d**). Red arrow highlights 50 kDa and blue arrow is 37 kDa

modify sulfhydryl groups on cysteines. Nitric oxide has been found to regulate membrane targeting and gap junction signaling in ECs. For example, eNOS has a direct effect on conducted vasodilation across the endothelium by regulating Cx37 and Cx40 trafficking and signaling, and S-nitrosylation of Cx43 regulates IP$_3$ transfer through myoendothelial gap junctions [7, 14, 15]. S-nitrosylation of cysteine residues is highly dependent on cysteine oxidation state and the amino acids surrounding these cysteines, thus not all cysteines on connexins are available for S-nitrosylation [16]. For example, the extracellular cysteines on Cx43 maintain protein structure through formation of disulphide bonds and are not thought to be S-nitrosylated. While the carboxyl terminus of Cx43 contains three seemingly available cysteines, but only 1 (C271) has been reported to be S-nitrosylated [14]. In order to identify S-nitrosylated proteins, a modified version of the biotin switch assay as first described by Jaffrey et.al. and Wang et al. can be employed (Fig. 5) [14, 17, 18].

In the previous chapter on immunoblotting, standard polyacrylamide gel electrophoresis was demonstrated. In this chapter, we will introduce modifications of this technique for identification of Cx43 phosphorylation in membrane preparations as a method for enhancing signal detection using phospho-specific antibodies and the biotin switch technique for the identification of Cx43 S-nitrosylation in cultured vascular cells. In addition, this chapter will discuss the use of fluorescent secondary antibodies for Western blotting which provides enhanced sensitivity and resolution for quantitative analysis over standard ECL detection.

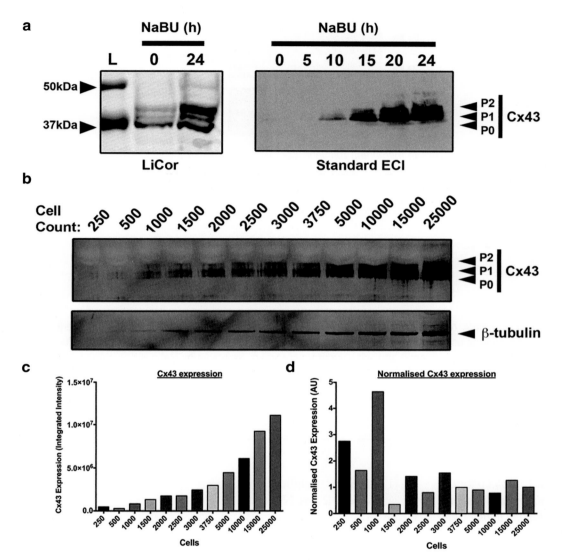

Fig. 3 (**a**) *Detection of Cx43 proteins by Western blotting using LI-COR Odyssey as compared to ECL detection.* In cell lysates from HeLa cells stably expressing Cx43 stimulated over a 24 h time course with sodium butyrate (NaBu 0.5 mM), good separation of the gel ensures proper visualization of the multiple phospho-isoforms of Cx43. Comparisons of LI-COR imaging and standard ECL demonstrate the upper limitations of detection for Cx43 by ECL (*right*) and enhanced sensitivity and lack of saturation in LI-COR (*Left*). (**b–d**) *The dynamic range of LI-COR imaging and quantification.* Specific numbers of HUVEC cells were counted and run by Western blot for the detection of Cx43 and β-tubulin. Cx43 signal is detected in as low as 250 cells and can be simultaneously viewed at the same time as 25,000 cells. It is worth noting that while the bands appear very dark on the blot, they are not saturated and can still be accurately quantified via LI-COR software (**c**). In (**d**), normalization of the Cx43 signal against β-tubulin demonstrates an increase in accuracy of signal detection above 1000 cells

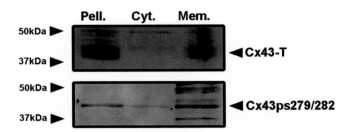

Fig. 4 *Detection of Cx43 in membrane fractions.* Comparisons between the expression of Cx43 (c-terminal antibody) and specific Cx43 phospho-isoforms using an antibody directed against the phosphorylated serines 279/282 (custom antibody from Dr Paul Lampe). Bands for the pS279/282 are primarily identified in the upper P2 band region

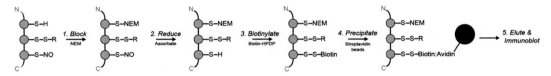

Fig. 5 *Schematic representation of the main stages of the biotin switch assay.* Following treatment and lysis of cells, NEM is used to block free thiols (step 1), following this *S*-nitrosylated residues are reduced by the addition of ascorbate (step 2) and subsequent free residues are bound by biotin HPDP (step 3). The resulting lysates can then be incubated with reactive streptavidin beads (step 4) to isolate the bound proteins, which can be removed from beads and run for Western blotting of the connexin proteins (step 5)

2   Materials

2.1   Cell Culture of Primary Vascular Human Cells

1. Human umbilical endothelial cells (HUVEC) purchased at passage 2 (*see* **Note 1**).

2. Human coronary artery smooth muscle cells (HCASMC) purchased at passage 2 (*see* **Note 1**).

3. Endothelial cell medium: medium 200, 20 % FBS, low serum growth kit (Thermo-Fisher Scientific) (*see* **Note 2**).

4. Smooth muscle cell medium: medium 231, 20 % FBS, smooth muscle growth serum (Thermo-Fisher Scientific).

5. Incubator at 37 °C, 5 % CO_2, 90 % humidity, and class 2 bio-safety cabinet.

6. Sterile cell culture six-well plastic plates.

7. 50 µM Fibronectin solution (*see* **Note 3**).

8. Trypsin–EDTA.

9. PBS: 137 mM NaCl, 2.7 mM KCl, 1.5 µM KH_2PO_4, 8.1 mM Na_2HPO_4 without divalent cations, e.g., $CaCl_2$ or $MgCl_2$.

2.2 Membrane Protein Isolation

1. Membrane isolation buffer 1: Ice cold D-PBS (137 mM NaCl, 2.6 mM KCl, 1.5 mM KH_2PO_4, 8.1 mM Na_2HPO_4, 1.2 mM $CaCl_2$, 1.05 mM $MgCl_2$), 1× protease inhibiter cocktail, 1× Phosphatase inhibitor cocktail (*see* **Note 4**), 10 mM sodium fluoride, 100 mM AEBSF (*see* **Note 5**).

2. Pellet lysis buffer: PBS, 1% SDS, 300 μM sodium orthovanadate (*see* **Note 6**), 100 mM DTT, 1× protease inhibitor cocktail, 1× phosphatase inhibitor cocktail, 100 mM AEBSF.

3. Dounce homogenizer (*see* **Note 7**).

4. Ice bucket/cold room.

5. Tube rotator.

6. Bench top micro-centrifuge cooled to 4 °C.

7. Ultracentrifuge, ultracentrifuge rotor, ultracentrifuge tubes cooled to 4 °C.

8. Sample loading buffer: 125 mM Tris–HCl pH 7.4, 40% glycerol, 1% SDS, 10% β-mercaptoethanol, 0.1% bromophenol blue.

2.3 Biotin Switch Analysis of Nitrosylated Proteins

1. Biotin switch lysis buffer: 250 mM HEPES pH 7.7, 1 mM EDTA, 0.1 mM neocuproine, 50 mM NaCl, 10 mM nethyl-maleimide (NEM), 1% NP-40, protease inhibitor cocktail (1:100), final pH (7.4).

2. 1 M HEPES pH 7.7: dissolve 283.3 g/L of HEPES in water and titrate pH to 7.7 with NaOH.

3. 25% SDS: Dissolve 100 g SDS into 400 mL dH_2O.

4. 1 mM neocuproine: Add 24.5 mg/100 mL in methanol.

5. HEN Buffer (nonreducing): 250 mM HEPES pH 7.7, 1 mM EDTA, 0.1 mM neocuproine.

6. HENS Buffer (reducing): 250 mM HEPES pH 7.7, 1 mM EDTA, 0.1 mM neocuproine, 1% SDS.

7. Biotin switch blocking buffer: To 9 volumes of HEN blocking buffer add 1 volume of 25% SDS and 20 mM NEM (*see* **Note 8**).

8. Biotin switch reducing buffer: PBS plus 1% SDS.

9. 100 mM sodium ascorbate (Vitamin C): Dissolve 0.198 g sodium ascorbate in 10 mL dH_2O.

10. 1 mM Copper II sulfate: Dilute 100 mM stock solution 1:100 in water to give a final concentration of 1 mM.

11. Biotin-HPDP: Biotin-HPDP-N-[6-(biotinamido)hexyl]-3'-(2'-pyridyldithio) propionamide. Prepare as a 50 mM stock in DMSO and store at –20 °C.

12. Streptavidin-coated agarose beads.

13. Opaque walled 1.5 mL eppendorfs (*see* **Note 9**).

2.4 Western Blotting

1. 4–12 % Bis–Tris precast polyacrylamide gels (*see* **Note 10**).

2. 20× MES running buffer (as earlier).

3. Gel running and transfer tanks.

4. Sample loading buffer.

5. Protein ladder.

6. Nitrocellulose.

7. Blotting paper.

8. Transfer buffer: 200 mM Glycine, 25 mM Tris Base, 20 % Methanol (no pH).

9. Ice.

2.5 Immunoblotting and Protein Detection Using LI-COR Odyssey

1. Blocking solution: Mix LI-COR blocking solution with 0.05 % tween-20 (*see* **Note 11**).

2. Primary antibodies.

3. Secondary antibodies with 680/700 and 800 nm fluorescent tags.

4. Rocking platform.

5. PBS-T: PBS containing 0.05 % tween-20.

6. LI-COR Odyssey scanner and software.

7. Stripping solution: 0.2 M NaOH pH 11–13.

3 Methods

3.1 Culture of Vascular Cells

1. To bring cells up, first prepare a cell culture flask 25 cm or 75 cm as required with media for the cells and place in the cell culture incubator for 20 min to acclimatize (*see* **Note 12**).

2. Thaw the vial of cells in a 37 °C water bath for approximately 2 min, until fully thawed.

3. Pipette cells into a centrifuge tube containing 5 mL of pre-warmed media and centrifuge at $300 \times g$ for 7 min (*see* **Note 13**). This step is used to remove the DMSO contained in the freezing serum for the cells.

4. Remove the media from the cell pellet and resuspend the cells using the media that is contained in the flask (prepared in **step 2**).

5. Cells should be allowed to settle for at least 24 h and used within appropriate passage numbers for the cells (*see* **Note 14**).

3.2 Setting Up Plates of Vascular Cells for Experiments

1. For HUVECs only, place 25 μL of fibronectin in the middle of each well of a six-well plate and use a cell scraper to spread the solution around the surface of the well.

2. Leave the plate in the cell culture hood to allow the fibronectin to dry in.

3. Once dry plates can be used immediately or sealed in plastic bags and stored at 4 °C for 1–2 weeks for later use.

4. At 80–90 % confluence, remove cell media from flasks and wash cells two times in 5 mL of PBS without calcium or magnesium, pipette in 750 μL of 1× Trypsin–EDTA. Rock the flask to cover the cells and place back in the incubator for 3 min. Check for cell dissociation by microscope. Once 80–90 % of cells have become dissociated add back 9 mL of media to cells. For continued passage add 2–3 mL of cell back to the flask and top up with 7 mL with fresh media.

5. Count the remaining cells using a hemocytometer and adjust the volume so that cells are approximately 5×10^4 per mL. This will ensure confluence in 1–2 days (*see* **Note 15**).

6. Place 1.5 mL of cells in each well of a six-well plate and place plate in incubator.

3.3 Membrane Preparations for Western Blotting of Phospho-Connexins

1. Grow cells to 70–90 % confluence in six-well plates.

2. Place plate on ice and if possible perform isolation in a cold room (*see* **Note 16**).

3. Rinse cells two times with ice-cold PBS 125 mM NaCl containing $CaCl_2$ and $MgCl_2$.

4. Add 1 mL ice-cold membrane isolation buffer.

5. Scrape cells from the plate surface using a cell scraper and pipette cell solution into prechilled 1.5 mL centrifuge tubes.

6. Using a dounce homogenizer, dounce samples in 30 times in tube on ice. This is required to break up cells, as there is no detergent in solution.

7. To remove large organelles, e.g., nuclei, endoplasmic reticulum, mitochondria, place sample tubes in a 4 °C micro-centrifuge and spin at $5000 \times g$ for 5 min.

8. Transfer the supernatant by pipette without disturbing the pellet to tubes suitable for ultracentrifugation. The pellet can be resuspended in 30 μL pellet lysis buffer and stored at −20 °C for later analysis.

9. Place ultracentrifuge tubes in a prechilled 4 °C ultracentrifuge rotor.

10. Place rotor in a 4 °C ultracentrifuge and spin at $100,000 \times g$ for 60 min.

11. Remove tubes from the ultracentrifuge and carefully remove all of the supernatant from the tube (*see* **Note 17**). The supernatant can be kept as a comparison of cytosolic fractions.

12. Resuspend pellet (membrane fraction) in 30 μL of pellet lysis buffer.

13. Add 5 μL of sample loading buffers and thoroughly mix by vortex,

14. Incubate in cold room with rotation for 30 min to ensure pellet dissolves fully.

15. Pulse centrifuge samples to ensure samples are at the bottom of the tubes.

16. Load 30 μL of sample onto precast 4–12 % Bis–Tris gels and run at 150 V until the loading dye front runs off of the gel. This ensures the best separation of the proteins from the phospo-protein bands.

17. Transfer proteins to nitrocellulose (*see* **Note 18**) by transfer in buffer for 1 h in cold room at 100 V.

18. Remove nitrocellulose to a clean container and wash 2× in PBS.

3.4 Detection of S-Nitrosylated Connexin Proteins Using the Biotin Switch Assay

1. Following treatment of cell with appropriate S-nitrosylating agent, e.g., 10 μM GSNO (*see* **Note 19**).

2. Isolate whole cell extract with lysis buffer (*see* **Note 9**).

3. Quantify protein concentration using BCA assay and use 300 μg protein for assay.

4. Precipitate protein by acetone precipitation using 2 volumes acetone. Incubate 30 min at –20 °C.

5. Spin samples in a micro-centrifuge for 5 min at $5000 \times g$ in 4 °C micro-centrifuge.

6. Remove supernatant and resuspend pellet in 100 μL HEN buffer.

7. Add 4 volumes (400 μL) blocking buffer. Incubate at 40 °C for 20 min with shaking.

8. Precipitate proteins using 2 volumes of acetone, incubate for 30 min at –20 °C, spin for 15 min in 4 °C micro-centrifuge at $5000 \times g$, remove and discard the supernatant.

9. Resuspend the pellet in 75 μL reducing buffer, add 1/3 volume (25 μL) of 4 mM Biotin-HPDP (1:12.5 dilution of stock) in DMSO made fresh.

10. Add 1 μL ascorbate (1 mM final conc.) and 1 μL Copper II sulfate (10 μM final) to each sample. Incubate for 1 h at room temperature.

11. Remove Biotin-HPDP by acetone precipitation using 2 volumes of acetone, incubate for 30 min at –20 °C, then spin for 15 min in cold micro-centrifuge at $5000 \times g$, remove and discard the supernatant.

12. Resuspend pellet in 100 μL HENS buffer and add 2 volumes of Neutralization Buffer (*see* **Note 20**).

13. Add 50 μL streptavidin–agarose beads to purify the biotinylated proteins. Incubate for 1 h at room temperature with rotating.

14. Wash five times with neutralization buffer plus 600 mM NaCl.

15. Elute proteins from the beads with 50 μL 2× SDS PAGE loading buffer and boil for 5 min.

16. Load 25 μL of protein onto a 4–12 % Bis–Tris gel and run at 150 V until the loading dye front runs off of the gel. This ensures the best separation of the proteins.

17. Transfer proteins to nitrocellulose (*see* **Note 18**) by transfer in buffer for 1 h in cold room at 100 V.

18. Remove nitrocellulose to a clean container and wash two times in PBS.

3.5 Detection of Proteins on Nitrocellulose Using LI-COR Odyssey

1. Block nitrocellulose membranes for 1 h in blocking solution (*see* **Note 21**).

2. Mix primary antibodies in blocking solution at the appropriate concentration (*see* **Note 22**).

3. Incubate with rocking overnight in cold room.

4. Remove antibody solutions from the blot, these can often be stored at −20 °C for reuse, depending on the antibody.

5. Wash blot 4 times for 15 min in PBS-T.

6. Mix the secondary antibodies, e.g., anti-mouse-700 and anti-rabbit-800 together at a concentration of 1:10,000 in blocking solution (*see* **Note 23**).

7. Incubate with rocking at room temperature for 1 h.

8. Remove secondary antibodies. These can be stored at −20 °C for later use.

9. Wash three times for 10 min in PBS-T.

10. Wash for 2 min in PBS (*see* **Note 24**).

11. Place blot face down on LI-COR Odyssey Scanner.

12. Perform initial low resolution scan to ensure settings for laser power and to ensure no saturation of signal.

13. Capture a final image in high resolution suitable for publication, e.g., 84/High settings.

14. Bands can be quantified using the built-in software from LI-COR, with automatic background detection.

15. For membrane proteins, GAPDH makes an appropriate loading control and can be used to normalize the protein signal. In S-nitrosylation experiments, total Cx43 from protein lysates harvested prior to the experiment can be used as loading controls for Cx43 expression (e.g., [14]).

16. Data from these can be exported to Excel and statistical analysis software for analysis of controls versus treatment (e.g., Fig. 3).

4   Notes

1. Primary HUVEC and HCASMC cells are supplied in frozen vials from the manufacturer and should be stored at −80 °C until use.

2. The low serum growth kit from Thermo-Scientific contains 1 mL gentamycin/amphotericin B (500×), a 1 mL mixture of basic fibroblast growth factor (1.5 µg/mL)/heparin (5 mg/mL)/BSA (100 µg/mL), 0.5 mL hydrocortisone (1 mg/mL), 1 mL EGF (5 µg/mL), and 5 mL of FBS. To make the EC media, we add all of the low serum growth kit with the exception of the FBS (which is not used) to the 500 mL of M200 media, mix for 10 min at room temperature, remove 100 mL and replace with 100 mL of FBS that has been batch tested for cell growth. The 100 mL of no-serum media has 50 µL of FBS added to make a final concentration of 0.1 % serum, which is useful for cell transfections and drug treatments and can sustain HUVEC for several days in culture without cell death.

3. Fibronectin coating of plates promotes adhesion and reduces the tendency of the endothelial cells to float off from the plate surface.

4. Protease and phosphatase inhibitor cocktails are 100× stock from Sigma. Sigma offers two separate phosphatase inhibitor cocktails. In our experiments both are used.

5. There are no detergents added to this buffer as this would lead to breakdown of the membrane components.

6. Initially, sodium orthavanadate will be in an inactive multimeric, e.g., decavanadate forms and has to be depolymerized to the active monomer form at pH 10. A 300 mM stock is made by dissolving in water and then adjusting to pH 10 using NaOH. The solution will turn yellow/orange and this should be boiled until clear and pH adjusted to 10. Repeat the process until the solution remains clear at pH 10. The stock can then be stored at −20 °C.

7. Use of a dounce homogenizer on ice reduces heat buildup in samples.

8. *N*-Ethylmaleimide (NEM) is an alkene that is reactive toward thiols and is commonly used to modify cysteine residues in proteins and peptides. This is also an important addition in pulling down proteins for coimmunoprecipitation.

9. S-nitrosylation modifications are a fairly labile and can be reduced by excessive light exposure. It is crucial that all portions of this protocol are carried out in opaque tubes that will protect the samples from light.

10. Bis–Tris composition gels are used as they are heat stable. This means that they can be stored at room temperature for up to 1 year and are less likely to warp during running of the gel.

11. It is generally recommended for phospho-proteins that a blocking solution is serum or BSA based (e.g., LI-COR blocking solution or Thermo Fisher SeaBlock) as opposed to dried milk solutions which contain casein which is a phospho-protein and can produce high background interfering with specific detection of the phospho-signal. This is not the case for all phospho-sites and should be tested as milk solutions are cheaper than the commercial blocking solutions.

12. All cell culture work and preparation should be performed in a Class II cell culture hood to avoid contamination of cells.

13. Using a gentle centrifuge speed to pellet the cells will result in a looser pellet that can be resuspended more easily causing less damage to the cells.

14. As both HUVEC and HCASMC are primary cells, long-term passage is not recommended. HUVEC cells doubling rate is around 24 h and is recommended for use within 16 population doublings and HCASMC cell doubling rate is around 48 h and is recommend for use within 16 divisions.

15. Optimally cells should never exceed 95 % confluence. Overconfluence should be avoided as cell phenotypes can alter and cells are more prone to peeling off in layers from the wells.

    For experiments that require a longer time course adjust cell concentration down, e.g., a 5 days experiment may require cells to be plated at 1.0×10^4 per mL.

16. All equipment and consumables used in the preparation should be cooled to 4 °C.

17. When setting the ultracentrifuge up, ensure to set deceleration speed to the minimum to ensure pellet does not loosen at the end of the cycle due to high braking forces.

18. Due to the sensitive nature of LI-COR Odyssey use 0.2 μm nitrocellulose. Use of 0.4 μm can lead to excessive background seen within the pores of the membrane. It is possible to use PVDF but ensure it is rated for fluorescent imaging as normal PVDF has a high background.

19. Controls for experiments will vary; examples of typical experimental controls for S-nitrosylated proteins, Cx43 proteins, and the appropriate controls can be found in the manuscripts by Jaffrey et al. and Straub et al. [14, 17].

20. At this point remove 20 μL of the lysate and add Laemmli buffer without reducing agents for use as loading controls.

21. No tween should be added in the first blocking step; this can be added prior to addition of the primary antibodies.

22. As the LI-COR can detect two channels, it is possible to use two simultaneous antibodies, e.g., Cx43 anti-rabbit and GAPDH anti-mouse. Antibodies should be checked separately first and then can be combined once the signal/banding pattern of each is established.

23. The signal to noise in the 800 channel is higher than in the 700 channel. As a result it is better to use stronger antibodies, e.g., GAPDH in the 700 channel and weaker antibodies, e.g., phospho-antibodies in the 800 channel.

24. Unlike ECL signal, the fluorescent antibody signal is very stable and can be stored well in PBS in a fridge or the membrane can be dried between filter paper and stored in the dark for analysis at a later date.

Acknowledgements

This work was financially supported by the University of Glasgow, Lord Kelvin Adam Smith Fellowship (SRJ), by NIH grants R00 HL11290402, the Institute for Transfusion Medicine and the Hemophilia Center of Western Pennsylvania (ACS) and postdoctoral fellowships from the Canadian Institutes for Health Research and Alberta Innovates - Health Solutions (AWL).

References

1. Johnstone S, Isakson B, Locke D (2009) Biological and biophysical properties of vascular connexin channels. Int Rev Cell Mol Biol 278:69–118

2. Straub AC, Johnstone SR, Heberlein KR et al (2010) Site-specific connexin phosphorylation is associated with reduced heterocellular communication between smooth muscle and endothelium. J Vasc Res 47:277–286

3. Johnstone SR, Kroncke BM, Straub AC et al (2012) MAPK phosphorylation of connexin 43 promotes binding of cyclin E and smooth muscle cell proliferation. Circ Res 111:201–211

4. Looft-Wilson RC, Billaud M, Johnstone SR et al (2012) Interaction between nitric oxide signaling and gap junctions: effects on vascular function. Biochim Biophys Acta 1818:1895–1902

5. Solan JL, Lampe PD (2007) Key connexin 43 phosphorylation events regulate the gap junction life cycle. J Membr Biol 217:35–41

6. Solan JL, Lampe PD (2014) Specific Cx43 phosphorylation events regulate gap junction turnover in vivo. FEBS Lett 588:1423–1429

7. Le Gal L, Alonso F, Mazzolai L et al (2015) Interplay between connexin40 and nitric oxide signaling during hypertension. Hypertension 65:910–915

8. D'hondt C, Iyyathurai J, Vinken M et al (2013) Regulation of connexin- and pannexin-based channels by post-translational modifications. Biol Cell 105:373–398

9. Nishi H, Hashimoto K, Panchenko AR (2011) Phosphorylation in protein-protein binding: effect on stability and function. Structure 19:1807–1815

10. Campbell AS, Johnstone SR, Baillie GS et al (2014) beta-Adrenergic modulation of myocardial conduction velocity: connexins vs. sodium current. J Mol Cell Cardiol 77:147–154

11. Johnstone SR, Ross J, Rizzo MJ et al (2009) Oxidized phospholipid species promote in vivo differential cx43 phosphorylation and vascular smooth muscle cell proliferation. Am J Pathol 175:916–924

12. Solan JL, Lampe PD (2009) Connexin43 phosphorylation: structural changes and biological effects. Biochem J 419:261–272

13. Berg AP, Talley EM, Manger JP et al (2004) Motoneurons express heteromeric TWIK-related acid-sensitive K+ (TASK) channels containing TASK-1 (KCNK3) and TASK-3 (KCNK9) subunits. J Neurosci 24:6693–6702

14. Straub AC, Billaud M, Johnstone SR et al (2011) Compartmentalized connexin 43 s-nitrosylation/ denitrosylation regulates heterocellular communication in the vessel wall. Arterioscler Thromb Vasc Biol 31:399–407

15. Meens MJ, Alonso F, Le Gal L et al (2015) Endothelial Connexin37 and Connexin40 participate in basal but not agonist-induced NO release. Cell Commun Signal 13:34

16. Tannenbaum SR, White FM (2006) Regulation and specificity of S-nitrosylation and denitrosylation. ACS Chem Biol 1:615–618

17. Jaffrey S.R., Snyder S.H. (2001) The biotin switch method for the detection of S-nitrosylated proteins. Sci STKE 2001:pl1

18. Wang X, Kettenhofen NJ, Shiva S et al (2008) Copper dependence of the biotin switch assay: modified assay for measuring cellular and blood nitrosated proteins. Free Radic Biol Med 44: 1362–1372

Chapter 8

Preparation of Gap Junctions in Membrane Microdomains for Immunoprecipitation and Mass Spectrometry Interactome Analysis

Stephanie Fowler, Mark Akins, and Steffany A.L. Bennett

Abstract

Protein interaction networks at gap junction plaques are increasingly implicated in a variety of intracellular signaling cascades. Identifying protein interactions of integral membrane proteins is a valuable tool for determining channel function. However, several technical challenges exist. Subcellular fractionation of the bait protein matrix is usually required to identify less abundant proteins in complex homogenates. Sufficient solvation of the lipid environment without perturbation of the protein interactome must also be achieved. The present chapter describes the flotation of light and heavy liver tissue membrane microdomains to facilitate the identification and analysis of endogenous gap junction proteins and includes technical notes for translation to other integral membrane proteins, tissues, or cell culture models. These procedures are valuable tools for the enrichment of gap junction membrane compartments and for the identification of gap junction signaling interactomes.

Key words Cx32, Endogenous immunoprecipitation, Gap junctions, Integral membrane protein, Interactome, Membrane fractionation, Membrane microdomains, Network analysis, Opti-Prep gradient, Proteomics

1 Introduction

Gap junctions were originally identified as intercellular membrane channels, yet are now understood to be complex signaling platforms with functional diversity extending far beyond metabolic coupling [1–3]. Gap junction plaques are composed of networks of integral scaffolding proteins with cytoskeletal contacts that recruit intracellular and nuclear localized enzymes and substrates (kinases, phosphatases, etc.) [4]. Sequestering multiprotein complexes at gap junction plaques is hypothesized to regulate the spatiotemporal proximity of signaling mediators, ensuring their proper subcellular localization for efficient signaling cascade generation [5, 6].

Most of our knowledge of gap junction protein interactions comes from studies of ubiquitously expressed Cx43 channels [6].

Mathieu Vinken and Scott R. Johnstone (eds.), *Gap Junction Protocols*, Methods in Molecular Biology, vol. 1437, DOI 10.1007/978-1-4939-3664-9_8, © Springer Science+Business Media New York 2016

There is significant diversity in connexin interactomes including calcium sensing proteins, other junction proteins, membrane channels, enzymes, cytoskeletal proteins, transcription factors, and proteins involved in intracellular trafficking [5]. Many of the interactions are shared amongst different connexin family members, and certain connexin isoforms can interact with each other in heteromeric and/or heterotypic configurations, thereby altering network interactions [7–9].

Connexin interaction data has been gathered using a variety of techniques, including colocalization imaging, immunoprecipitation (IP) assays, affinity binding assays, and biochemical techniques [2]. Limitations common to many of these techniques include high false-positive (and false-negative) interactions, inability to detect transient/weak interactions, and masking of less abundant proteins by highly expressed proteins. Spurious interactions often occur during the mixing of cellular compartments upon cell lysis, generating nonphysiological interaction environments [10, 11]. Further, the use of ectopic epitope-tagged proteins may interfere with localization/posttranslational modification of bait proteins while the tag itself may mask sites of protein–protein interaction, again generating nonphysiological interactions [12, 13]. Despite these limitations, *in vivo* systems are usually preferred over synthetic systems for multiprotein complex detection, as proteins in synthetic systems may not assemble or traffic properly, and crucial accessory proteins may not be present to enable full complex assembly [14].

To address these technical limitations, the present work describes a procedure for fractionating both light and heavy cellular membranes from tissue using detergent-free OptiPrep™ density gradients to identify microdomains enriched for any membrane protein of interest. We identify here the membrane microdomains that are enriched for the liver gap junction protein Cx32 and describe how to analyze endogenous Cx32-associated interaction proteins following gentle membrane solvation, IP, and tandem mass spectrometry (MS/MS). Technical guidance for interaction protein validation and network analysis is included, and options for examining integral membrane protein interactions in other tissues and cell culture models are discussed.

2 Materials

2.1 Mouse Liver Tissues

1. Mice: wild-type (WT) and knockout (KO, negative control) animals for dissection (*see* **Note 1**).

2. For identification of Cx32 complexes: 1–3 g of isolated mouse liver material from 2 to 3 mice per centrifuge tube (*see* **Notes 2 and 3**).

3. For the identification of less abundant PM microdomains more starting material is required, e.g., 8–10 g liver wet weight recommended (*see* **Note 4**).

2.2 Tissue Homogenization and OptiPrep Membrane Fractionation

1. 10 mM phosphate buffered saline (PBS): 10 mM phosphate, 137 mM NaCl, pH 7.2, filter sterilized.

2. Light membrane homogenization buffer (LHB): 0.25 M sucrose, 2 mM $MgCl_2$, 20 mM HEPES-NaOH (*see* **Note 5**). Adjust pH to 7.4 and add 1 mM sodium fluoride, 50 μg/mL aprotinin, 1 mM sodium orthovanadate, 1 mg/mL PMSF immediately *prior* to use. Make 500 mL buffer just *prior* to use, filter sterilize, and keep on ice.

3. Heavy membrane homogenization buffer (HHB). 250 mM sucrose, 1 mM EDTA, 20 mM HEPES-NaOH (*see* **Note 6**). Adjust pH to 7.4 and add 1 mM sodium fluoride, 50 μg/mL aprotinin, 1 mM sodium orthovanadate, 1 mg/mL PMSF immediately *prior* to use. Make 500 mL buffer just *prior* to use, filter sterilize, and keep on ice.

4. Potter-Elvehjem glass homogenizer with loose and tight fitting teflon pestles (30 mL).

5. 15 and 50 mL tubes.

6. Glass pasteur pipettes.

7. OptiPrep (OP) density gradient solution (Sigma).

8. OP Diluent: 0.25 M sucrose, 120 mM HEPES, pH to 7.4 with NaOH. Make 50 mL and filter sterilize.

9. 50 % working solution of OP in LHB or HHB: 5 parts OP with 1 part OP diluent.

10. Protein concentrating centrifuge tubes, e.g., Amicon Ultra-4 centrifugal unit.

11. 10 % detergent of choice in LHB and HHB, e.g., triton X-100, NP-40, digitonin, CHAPS (*see* **Notes 7** and **27**).

12. Refrigerated high-speed centrifuge with swinging bucket rotor fitting 15 mL tubes (capable of at least $3000 \times g$).

13. Rotor for high-speed centrifuge fitting 50 mL polypropylene bottles with screw-on caps.

14. Ultracentrifuge with swinging bucket rotor for 13 mL tubes.

15. 13.2 mL (14×89 mm) polypropylene ultracentrifuge tubes for swinging bucket rotor.

2.3 Agarose Bead Preparation

1. Protein G and/or Protein A agarose beads (*see* **Note 8**).

2. Monoclonal or polyclonal antibody directed against connexin/membrane protein of interest. These should be validated for use in immunoprecipitation studies.

3. 10 mM PBS: 10 mM phosphate, 137 mM NaCl, pH 7.2, filter sterilized.

4. Tube rotator.

5. Refrigerated tabletop centrifuge.

2.4 Immuno-precipitation (IP) Complex Preparation for MS/MS

1. IP buffer 1: 20 mM Tris, 137 mM NaCl, 2 mM EDTA, 1% NP-40, pH 7.4. Make 50 mL just before use and store up to 1 week at 4 °C.

2. IP buffer 2 (no detergent) : 20 mM Tris, 137 mM NaCl, 2 mM EDTA, pH 7.4. Make 50 mL just before use and store up to 1 week at 4 °C.

3. Ammonium hydroxide elution buffer: 0.5 M NH_4OH, 0.5 mM EDTA.

4. Centrifugal vacuum concentrator, e.g., speedvac (Thermo Scientific).

5. 6× Sample buffer: 350 mM Tris–HCl, pH 6.8, 5% glycerol, 10% SDS, 100 mM DTT, 0.002% bromophenol blue.

6. 2× SDS sample buffer: 6× sample buffer diluted to 2× with buffer containing 350 mM Tris–HCl, 0.28% SDS, pH 8 buffer.

7. β-Mercaptoethanol (BME).

8. Precast 4–12% Bis–Tris gels gradient gels.

9. Antibodies for fraction characterization: Cx32, Flotillin, $Na^+K^+ATPase$, Calnexin, Golgin97, LAMP1, CoxIV.

2.5 Silver Staining and Sample Preparation for MS/MS

1. Clean, acid washed glassware rinsed with at least three changes of double distilled water (ddH_2O).

2. 5% Acetic acid/methanol solution: 25 mL acetic acid in 475 mL ddH_2O and 500 mL methanol.

3. 0.02% Sodium thiosulfate: 0.2 g sodium thiosulfate in 1 L ddH_2O.

4. 0.1% Silver nitrate solution: 0.1 g $AnNO_3$ in 100 mL ddH_2O.

5. 0.01 Formaldehyde in 2% sodium carbonate: 150 μL 37% formaldehyde and 10 g sodium carbonate in 500 mL ddH_2O.

6. 1% Acetic acid: 10 mL acetic acid in 990 mL ddH_2O.

7. #11 scalpel blades.

8. MS/MS Facility of choice.

2.6 Considerations for Reducing Contamination in Downstream MS/MS Analyses (See Note 9)

1. Nonpowdered, nitrile gloves cleaned with 70% ethanol. Gloves should be changed often.

2. A clean lab coat and low-shedding clothing, e.g., no wool.

3. Nonautoclaved, filtered pipette tips.

4. Laminar flow hood or a keratin-free room if possible.

5. Low-bind eppendorf tubes.

6. Clean, acid washed glassware rinsed with at least three changes of ddH_2O.

7. Noncontaminated cell cultures.

8. Filter sterilized solutions (*see* **Note 10**).

9. Clean and sterile workspace (*see* **Note 11**).

2.7   Validation and Analysis of Protein Interactions

1. Network analysis software, e.g., ingenuity systems pathway analysis software, version 8.8, Ingenuity Systems Inc.

3   Methods

3.1   Tissue Homogenization and OP Membrane Fractionation (See Notes 12 and 13)

1. Prepare WT and KO animals for dissection.

2. Carefully excise the gallbladder (filled with digestive enzymes), remove the livers, and immediately place into ice-cold LHB (*see* **Notes 14** and **15**). Recommended starting material is 2–3 livers (3 g total wet weight) per genotype.

3. Mince tissue with a razor blade on a teflon chopping board until homogenous pieces of around 2 mm² are achieved.

4. Wash liver pieces with three changes of 30 mL ice-cold LHB in a 50 mL tube. At each wash, invert the liver pieces in the fresh wash solution, allow the liver pieces to settle at the bottom of the tube, and pour off the supernatant.

5. Discard the final wash supernatant and add LHB to the liver pieces at 4× the volume per wet weight of tissue, e.g., 12 mL LHB per 3 g liver tissue.

6. Swirl the tube to suspend the liver pieces in LHB and pour the entire contents into a 30 mL glass homogenizer. Transfer any leftover liver pieces to the homogenizer using forceps.

7. Homogenize gently on ice until very smooth. This will take about 5 strokes with a loose-fitting pestle and 20–30 strokes with a tight-fitting teflon pestle (*see* **Notes 16** and **17**).

8. Transfer homogenized lysate to 15 mL falcon tubes and centrifuge at $800 \times g$ for 5 min at 4 °C to remove nuclei and cellular debris.

9. Collect the supernatant and repeat centrifugation at $800 \times g$ to collect the final postnuclear supernatant (PNS).

10. Reserve a 100 μL aliquot of the PNS at −80 °C for later analysis.

11. Transfer the PNS to polypropylene centrifuge bottles for high speed centrifuge, and centrifuge supernatants at $8000 \times g$ for 10 min to pellet intact mitochondria. Reserve pellet on ice.

12. Remove the supernatant to a clean bottle and repeat centrifugation at $8000 \times g$ for 10 min to remove the majority of intact mitochondria. Reserve pellet on ice.

13. Carefully transfer the postmitochondrial supernatants (PMS) into prechilled 15 mL tubes (*see* **Note 18**).

14. Reserve a 100 μL aliquot of the PMS at −80 °C.

15. Concentrate the PMS down to 1.4 mL using a centrifugal unit with 3 kDa molecular weight cutoff.

16. Centrifuge as recommended by the manufacturer at 4 °C for 30 min (*see* **Note 19**).

17. Prepare the HM pellet (mitochondrial and mitochondrial membranes) and dilution of OP solutions while the PMS is concentrating.

18. To prepare the heavy membranes for flotation, gently combine the mitochondrial pellets stored on ice with 2 mL HHB using a 1 mL pipette with the tip cut off.

19. Add 28 mL HHB to the resuspended pellets and centrifuge at 4 °C for 10 min at 8000 × *g*.

20. Discard the supernatant and carefully resuspend the pellet a second time in 2 mL HHB.

21. Add 28 mL HHB and pellet the membranes a final time for 4 °C for 10 min at 8000 × *g*.

22. Resuspend the heavy membrane pellet (HMP) very carefully in 250 μL HHB using a 1 mL pipette with the tip cut off. The final volume of the resuspended pellet should be 400–500 μL (*see* **Note 18**).

23. Reserve a 50 μL aliquot of the resuspended HMP at −80 °C for later analysis.

24. During preparation of the HMP for flotation and while concentrating the PMS, prepare appropriate volumes of OP dilutions from the 50 % OP working solution as required. Dilute 50 % OP with LHB or HHB to prepare the appropriate volumes of OP dilutions for the LM and HM flotations (LM: 25, 22.5, 20, 15, and 10 %; HM: 30, 26, 24, 22, and 20 %) as shown in Fig. 1. Chill all solutions on ice.

25. Following concentration of the PMS, reserve a 100 μL aliquot at −80 °C for later analysis.

26. Dilute 1.24 mL of the concentrated PMS into 1.86 mL 50 % OP working solution to generate 3.1 mL of 30 % PMS/OP solution. Mix well by inversion.

27. Dilute 372 μL of the HMP into 2.728 mL 50 % OP working solution to generate 3.1 mL of 44 % HM/OP solution. Mix well by inversion.

28. Add 3 mL of the 30 % PMS/OP solution and 3 mL of the 44 % HM/OP solution to separate chilled 13.2 mL polyallomer centrifuge tubes on ice, ensuring to add each solution to the bottom of the tube without touching the sides.

29. Carefully overlay the appropriate volume and concentration of OP dilution solutions on top of the 30 % PMS/OP and 44 % HM/OP layers by placing the tip of a 1 mL pipette at the interface of the bottom layer and slowly drawing it up the side of the tube as the lighter density solution is released.

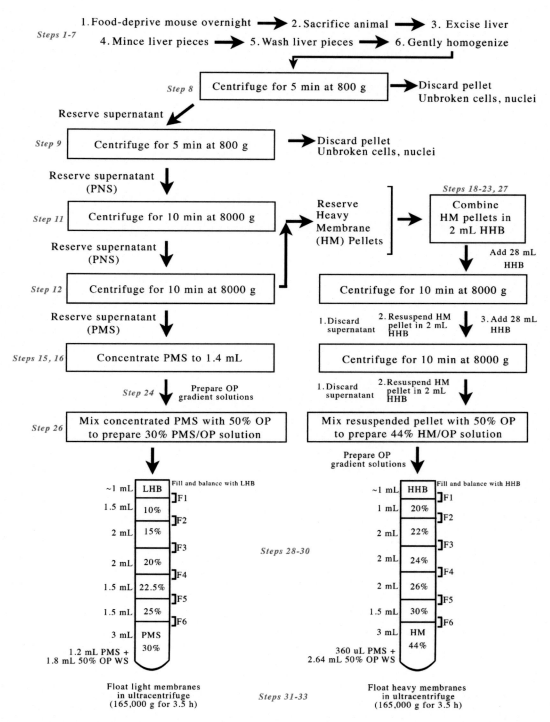

Fig. 1 *Membrane fractionation experimental flow-through.* Briefly, whole liver tissue is separated into light membrane (LM) and heavy membrane (HM) fractions by differential centrifugation. Nuclei are cleared following two centrifugations at $800 \times g$, and HM and mitochondria are cleared from the PNS following two centrifugations at $8000 \times g$. LM and HM are floated in 30%-0%, and 44%-0% OP gradients, respectively, to generate six different density microdomains per tube. The LM preparation contains the majority of plasma-membrane (PM)-associated gap junction proteins, concentrated in the 20-0% OP layers. The HM may be floated to collect the PM and lipid raft domains from the top two interfaces, and to isolate mitochondria and lysosomes for further downstream processing

30. Continue overlaying the remaining density solutions as shown in Fig. 1, making sure that each layer sits discretely on top of the previous layer (*see* **Notes 20** and **21**).

31. Place filled tubes into chilled swinging buckets using a vertical support (*see* Fig. 2a, b), weigh the assemblies in matching pairs (±10 mg) with the opposite tube and bucket using LHB or HHB.

32. Tighten the lids onto the swinging buckets and attach each bucket to the rotor.

33. Centrifuge overnight at $165,000 \times g$ at 4 °C for 3.5 h (*see* **Note 22**).

3.2 Agarose Bead Preparation (See Note 23)

1. To reduce shear forces on the agarose beads, cut off the first few mm of the plastic pipette tip using a razor blade sterilized with 70 % ethanol.

2. Prepare protein G or protein A agarose beads by rinsing 1 mL bead slurry (50 % beads, 50 % PBS) with 10 volumes of PBS (*see* **Note 8**).

3. Invert to mix and pellet beads at $1000 \times g$ for 5–10 s (discard wash supernatant) and repeat wash to fully remove bead preservative.

4. After final wash, pellet beads at $1000 \times g$ for 5–10 s and discard supernatant. Regenerate the original bead volume (1 mL) by adding half the bead volume of PBS.

5. Add between 20 and 50 µg of a primary antibody of choice per 1 mL of beads (*see* **Note 24**) and rotate with inversion overnight at 4 °C (*see* **Notes 25** and **26**).

3.3 Fraction Collection and Preparation for IP

1. Carefully remove centrifuge tubes from the swinging buckets and set tubes in vertical stands. Membrane domains float to the interfaces between the OP layers (Fig. 2c, d).

2. Using a 1 mL pipette, collect the interfaces in 0.5–1.5 mL volumes by pipetting from the top of the tube.

3. Transfer the interfaces to chilled 1.5 mL eppendorf tubes.

4. Collect identical fraction volumes from WT and KO fractionations.

5. At this point samples can be either stored frozen at −80 °C for immunoblotting or maintained chilled for IP.

3.4 Characterizing Samples by Immunoblotting

1. Thaw samples and add nonionic detergent of choice to a final concentration of 1 % to each fraction (*see* **Note 27**).

2. Incubate fractions for 30 min at 4 °C with inversion rotation.

3. Centrifuge fractions at $16,100 \times g$ in a refrigerated tabletop centrifuge for 20 min at 4 °C to clear the lysates of any unsolubilized material.

4. Pipette the supernatants into clean tubes.

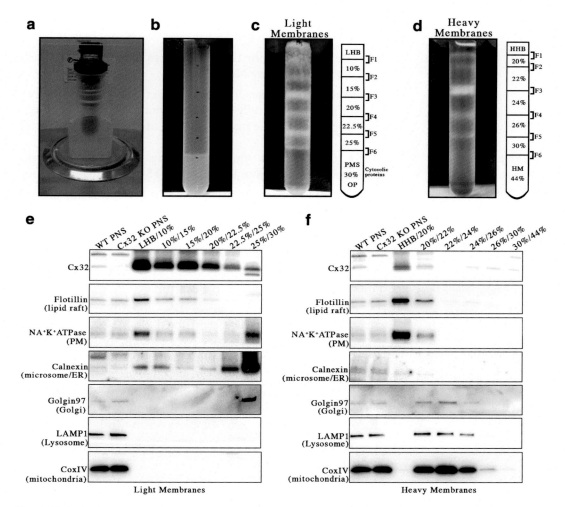

Fig. 2 *Typical set up and results from membrane fractionation.* It is critical to properly balance samples for ultracentrifugation to ±10 mg. To balance swinging buckets, place the swinging bucket, filled centrifuge tube, and lid into a plastic jar (or similar device) to balance it with an opposing bucket (**a**). An example of a fractionation tube with bottom-loaded LM sample, just prior to centrifugation with gradient interfaces marked (**b**). In LM fractionation, LM are light in color, almost pure white in the lightest fractions and cytosolic proteins are found in the 30 % OP layer at the bottom of the LM flotation (**c**). In HM fractionation, HM domains are darker in color compared to the LM domains, and the majority of the material bands at the 22/24 % interface (**d**). Western blot analysis of LM preparations containing gap junction proteins (Cx32), lipid rafts (Flotillin), plasma membranes (Na⁺K⁺ATPase), microsomes/ER (Calnexin), the majority of golgi membranes (Golgin97), and do not contain lysosomes (LAMP1) or mitochondria (CoxIV) (**e**). Western blot analysis of HM preparations containing gap junction proteins, lipid rafts, and plasma membranes in the lightest two fractions (24-0 % OP). The majority of the material is mitochondrial and located between the 26–20 % OP layers (**d**). Golgi membranes and lysosomes cofractionate with the mitochondria

5. Assay each fraction for protein concentration using a detergent compatible protein assay kit. Use a BSA standard made up in the same detergent buffer as the fractions.

6. Prepare an immunoblot of each solubilized fraction and include a sample of the original PNS solution (*see* **Note 28**).

7. Blot the membrane for the bait protein and several organelle markers (optional) (*see* **Note 29**).

8. From the immunoblots, identify all fractions in which the bait protein is found, or only fractions containing the bait protein and a certain organelle markers for use in IP studies (Fig. 2e, f).

3.5 Preparation of Samples for Immunoprecipitation

1. Based on previous immunoblotting identification, pool together the fractions of interest into 15 mL centrifuge tubes.

2. Add nonionic detergent of choice to a final concentration of 1% to the pooled fractions.

3. Incubate samples with inversion rotation for 30 min at 4 °C.

4. Centrifuge fractions at $16,100 \times g$ in a refrigerated centrifuge for 20 min to clear the lysates of any unsolubilized material.

5. Pipette the supernatants into clean tubes.

6. To exchange the buffer to an IP-compatible buffer, place supernatants in a 3 kDa centrifugal device and centrifuge according to the manufacturer recommendations at 4 °C to reduce the volume to a minimum of 50 µL (centrifugation time depends on total volume).

7. Discard the flow through, add 3.95 mL IP buffer 1 to the filter device and repeat the above centrifugation. Add IP buffer 1 to the concentrated sample to generate ~5 mg/mL lysate.

8. Reserve the final sample in a clean tube on ice (*see* **Notes 30** and **31**).

3.6 Immuno-precipitation Complex Preparation for MS/MS

1. To preclear the lysates, prepare 100 µL bead slurry per 1 mL of lysate.

2. Rinse the bead slurry with 10 volumes of PBS, invert to mix, and centrifuge at $1000 \times g$ for 5–10 s.

3. Discard the supernatant, add 10 more volumes of PBS, and resuspend beads by inversion.

4. Divide the resuspended bead slurry into two tubes (for WT and KO lysates) and centrifuge each bead tube at $1000 \times g$ for 5–10 s.

5. Remove the PBS, leaving the bead matrices only.

6. Immediately add the pooled fraction lysates to the rinsed bead pellets (*see* **Note 32**).

7. Rotate with inversion for 1 h at 4 °C (*see* **Note 33**).

8. Centrifuge the reactions at $1000 \times g$ for 10 s and pipette the supernatants to clean tubes.

9. Centrifuge supernatants for 1 min at $1000 \times g$ to fully clear the lysates of beads and pipette the precleared pooled fraction lysates into clean tubes.

10. Recover the previously prepared antibody coated beads and fully resuspend by gently tapping and inverting the tubes.

11. Transfer equal volumes of bead slurry to separate tubes for the WT and KO lysates.

12. Let beads settle by gravity and discard the bead supernatants.

13. Immediately add precleared pooled fraction lysates to each bead matrix and incubate overnight at 4 °C with inversion rotation.

14. Centrifuge samples at $1000 \times g$ for 10 s.

15. Remove supernatant, aliquot into 1 mL volumes, and store at −80 °C for later analysis as the post-IP lysate.

16. Carefully pipette 1 mL PBS onto the beads and invert gently to mix.

17. Centrifuge the beads at $1000 \times g$ for 5–10 s and discard the supernatant. Repeat the bead washing procedure 2 times for a total of 3 washes. Completely remove the final wash supernatant (*see* **Note 34**).

18. Elute protein complexes from the beads by adding 1 bead volume of ammonium hydroxide elution buffer (e.g., 200 µL for 200 µL bead slurry). Invert the beads to mix and centrifuge at $1000 \times g$ for 5–10 s. Repeat this process while collecting and pooling the elution volumes until <1 mL of ammonium hydroxide buffer is collected. Perform a final centrifugation at $1000 \times g$ for 1 min to ensure no beads remain in the final elution.

19. Speedvac this solution to lyophilize the protein complexes (*see* **Note 35**).

20. Reconstitute the lyophilized protein in 60 µL 2× SDS sample buffer containing 10 % BME. Boil for 5 min, or vortex well and incubate at RT for 30 min *prior* to loading on a gel.

21. Load up to 50 µL of eluted proteins onto a precast 4–12 % gradient gel (*see* **Note 36**).

22. Prepare and load 0.5 µL of unstained protein ladder (according to manufacturer's specifications) on each end of the gel.

23. Leave 1 lane space in between each WT and KO sample and between each ladder and the samples. Pipette an equal volume of sample buffer in empty gel lanes.

24. Do not run the dye front off the gel, as potential interacting partners could be lost.

25. Add approximately 100 mL of 5 % acetic acid/methanol fixative solution to a clean glass dish with lid and transfer the gel into the fixative immediately following electrophoresis.

26. Cover the glass container and incubate with gentle shaking in fixative for 30 min.

27. Rinse twice with ddH_2O for 2 min each.

28. To minimize background, wash gel overnight in ddH_2O with shaking at 4 °C.

3.7 Silver Staining and Sample Preparation for MS/MS (Day 4)

1. Sensitize the gel for 2 min in approximately 100 mL of 0.02 % sodium thiosulfate solution and rinse twice for 30 s each with ddH_2O.

2. Incubate the gel in approximately 100 mL fresh 0.1 % silver nitrate solution for 30 min.

3. Wash gel twice in ddH_2O for 30 s.

4. Develop the gel with two changes of freshly prepared 0.01 % formaldehyde in 2 % sodium carbonate solution. The first application of sodium carbonate solution will change to a brownish-yellow color after a few minutes of shaking. Discard this solution and replace with fresh sodium carbonate. During this second wash, the gel will develop color more quickly, so watch carefully for the appearance of bands. Do not overdevelop the gel, as this impairs the ability to identify proteins within the gel bands by MS/MS.

5. When the gel is sufficiently stained, discard the developing solution and add 1 % acetic acid to stop the reaction. The gel may be stored in 1 % acetic acid in the fridge for several weeks. A representative silver stained gel image can be found in [7] (*see* **Note 37**).

6. Wash new plastic overhead sheets with 70 % ethanol and sandwich the stained gel between the 2 sheets.

7. Image the gel and determine which bands will be analyzed.

8. Excise protein bands with a #11 scalpel blade (*see* **Note 38**).

9. Excise equivalent regions of each WT and KO gel lanes, cut each piece into smaller pieces approximately 3 mm² and transfer to separate 0.6 mL eppendorf tubes of mass spectrometry quality.

10. Add a region up to 25 kDa to each tube and continue this process until all regions of interest are collected from both genotypes.

11. Send samples for MS/MS analyses to a preferred MS facility. Ensure that this sample preparation will meet the facility's particular sample specifications.

3.8 Validation and Analysis of Protein Interactions

1. Repeat experiment to perform mass spectrometry on replicate IPs (minimum 2, preferably 3 replicates) (*see* **Note 39**).

2. Prioritize the validation of both previously identified protein interactions and novel interactions using a network analysis software package.

3. Following prioritization of protein interactions, perform reciprocal coimmunoprecipitations (co-IPs) to confirm that bait proteins and MS-identified proteins are present in the same complex (*see* **Note 40**). Representative reciprocal co-IP images can be found in [7].

4. To begin the co-IP, incubate agarose beads with primary antibodies overnight (*see* **Note 41**).

5. Following overnight bead coupling, homogenize liver tissue using IP buffer 2 with fresh protease inhibitors at 4× the tissue weight in a glass homogenizer (*see* **Note 42**).

6. To generate the PNS, centrifuge homogenate at $800 \times g$ for 5 min at 4 °C two times to clear unbroken cells and nuclei.

7. If mitochondrial protein interactions are not of interest, generate the PMS by centrifuging the supernatant at $8000 \times g$ for 10 min at 4 °C 2×, discarding the mitochondrial pellet each time.

8. Add nonionic detergent at the same concentration used to solubilize the membrane fractions in the original experiment.

9. Incubate samples with inversion rotation for 30 min at 4 °C.

10. Centrifuge at $16,000 \times g$ for 20 min at 4 °C. Reserve supernatants and adjust to 5 mg/mL protein with fresh IP buffer 1, following protein assay using a detergent compatible kit.

11. Add between 250 and 1000 μg of total protein for each co-IP to antibody/bead slurry and incubate with inversion at 4 °C overnight.

12. Centrifuge beads at $1000 \times g$ for 1 min.

13. Remove supernatant, aliquot into 1 mL volumes, and store at −80 °C for later analysis as the post-IP lysate.

14. Carefully pipette 1 mL PBS onto the beads and invert gently to mix.

15. Centrifuge the beads at $1000 \times g$ for 5–10 s and discard the supernatant. Repeat the bead washing procedure two times for a total of three washes. Completely remove the final wash supernatant.

16. Elute the protein complexes in two bead volumes (50–100 μL) 2× SDS sample buffer + 10 % BME.

17. Perform co-IP immunoblots, including the pre- and post-IP PNS samples, at least 1 KO sample and each of the experimental IP samples.

18. To determine success of the IP, blot for the bait protein using an antibody raised in a different animal, to avoid detection of the IgG signal.

19. Strip the blot and reprobe for potential interacting proteins. An antibody of a different species from the IP antibody is also preferred.

20. Revisit the network analysis when you have validated (or "un-validated") several of the MS-identified proteins to generate new hypotheses about the protein of interest.

4 Notes

1. If heavier membranes from mitochondrial pellets are to be analyzed, food-deprive the animals overnight to deplete the liver of glycogen. Glycogen cofractionates with mitochondria and will prevent the isolation of pure mitochondrial-associated membrane fractions [15].

2. A single liver from an adult mouse of around 25–30 g has a wet weight of approximately 1–1.5 g. Thus, 2–3 adult mouse livers per centrifuge tube are required for the identification of Cx32-based interactions.

3. This method is applicable to other tissues but will require optimization of the tissue amount required.

4. For more than three livers, set up identical centrifuge flotations using 13 mL tubes, or scale up the flotation, e.g., to accommodate up to 9 g wet weight liver using 38.5 mL tubes in a larger swinging bucket rotor.

5. Including $MgCl_2$ in the LHB helps in the isolation of intact plasma membrane sheets. Always use LHB when working with any of the light membrane supernatants or fractions *prior* to detergent solubilization.

6. HHB contains EDTA to chelate cations, as mitochondria are uncoupled in the presence of $Ca2^+$. Always use HHB when working with any of the heavy membrane pellets or fractions *prior* to detergent solubilization.

7. Detergent choice will be based on the proteins requiring isolation in each particular fraction, e.g., Triton X-100 insoluble lipid raft proteins from the lightest interfaces in each tube (Fig. 2).

8. Antibody/Agarose bead compatibilities. Protein G and A agarose beads bind strongly to most human and mouse IgG proteins; however, Protein A is usually preferred for binding rabbit IgG proteins. Some IgG subtypes do not bind strongly to Protein A or Protein G.

9. Protein identities resulting from MS/MS analyses rely heavily on the quality of the sample being investigated. MS/MS identifies the most abundant species in a sample and if low abundance species are of interest, it is especially important to limit sample contamination. Common contamination sources in MS analyses include bacteria from infected cell culture samples,

oils, plastics, fibers, dust, skin constituents from instruments and human handlers, and detergent/chemical residues used to clean laboratory equipment. For these reasons, always use good laboratory practice.

10. Prepare all solutions and buffers fresh and maintain sterility of stock solutions throughout the procedure. Keep lids closed and only open in a laminar flow hood.

11. The workspace and pipettes should be cleaned often with 70% ethanol and low-lint wipes.

12. Pilot experiments should be performed to characterize the fractions and determine which membrane domains contain your protein of interest and/or organelle marker of interest. These samples should not be used for the IP as they will have to be frozen to allow time for fraction analysis by immunoblotting. Freeze-thawing the samples may disrupt protein complexes. For the pilot experiment, omit the agarose bead preparation.

13. The complete procedure requires 4 consecutive lab days to generate a sample for MS/MS. Some of the different sections of the protocol require to be run continuously, e.g., antibody coating of beads on day 1 for use on day 2.

14. KO tissue is required for these studies.

15. Keep all samples on ice or at 4 °C at all times.

16. It is critical to homogenize gently with fluid motions and to prevent the introduction of air bubbles. Protection of organelle integrity during homogenization is essential for isolation of pure membrane fractions.

17. Modifications for cell culture and other tissues. Cell culture: remove cells from incubator and place on ice. Wash cells three times with ice-cold PBS and aspirate fully between each wash to remove all traces of serum. Add 1 mL cold LHB with protease inhibitors and scrape cells from the plate (Use 1 mL for every 5 plates scraped). Add this lysate to subsequent plates and scrape to pool lysates. Pass cells 8–10 times through a 23–25 g needle to break open cells instead of using a glass homogenizer and continue to PNS isolation. In certain cases of cells or tissue containing low amounts of mitochondria, the entire PNS supernatant may be fractionated in the light membrane schematic. Other tissues: Harder tissues such as heart and muscle will require more strokes in the glass homogenizer to achieve adequate disruption. Rotary blade homogenization may cause cross-contamination of subcellular fractions [16].

18. The membranes contained within PMS and HM pellets can be purified from soluble cytosolic proteins by ultracentrifugation at $100,000 \times g$ for 50 min at 4 °C following 10× dilution with LHB or HHB, preflotation. Discard the supernatant containing

cytosolic proteins and resuspend light and heavy membrane pellets in 3 mL of 30 % or 44 % OP solutions, respectively, and continue to fraction collection. Membrane fractions collected postflotation may also be purified by centrifugation at $100,000 \times g$ for 50 min at 4 °C following a 10× dilution in LHB or HHB.

19. Centrifugal filters have various starting volumes. A 4–15 mL volume is recommended for this technique. 3 g liver tissue will result in around 8 mL of PMS, which is concentrated 6× to achieve a final volume of approximately 1.4 mL. Centrifugation times will vary depending on the preparation, so this should be continually monitored to prevent drying of the filter membrane.

20. Sharp interfaces will make the membranes easier to collect.

21. Top up the centrifuge tube to a point 1 mm below the top edge with LHB, as tubes have no caps, they will warp if not completely filled.

22. It is critical that the layers are not disturbed by sudden acceleration/deceleration so these must be set to a minimum, e.g., deceleration = no brake/coast.

23. This step can be performed at any point on day 1.

24. Antibody concentration will depend on the quality of the antibody, how well it detects antigens in the IP environment, and expression level of the bait protein. An appropriate antibody:bead ratio will have to be empirically determined for each antibody.

25. As an extra negative control for MS/MS prepare beads coated with the same concentration of an isotype-specific IgG antibody.

26. If the bait protein is very close in size to 25 kDa (light chain IgG) or 50 kDa (heavy chain IgG), its identification may be masked on a silver stained gel by the eluted immunoglobulins. DMP can be used to chemically couple the antibodies to the beads reducing this background signal. Following overnight incubation with the primary antibody, cross-link the antibodies to the beads as described in [7].

27. An appropriate detergent will have to be experimentally determined for your membrane protein and tissue of interest. Nonionic detergents such as NP-40 and Triton X-100 are the most gentle, and at 1–2 % will solubilize most membrane domains while maintaining protein structure and protein–protein interactions. Digitonin is another mild nonionic detergent, very effective at solubilizing membrane proteins that should be evaluated during the optimization stage of IP or co-IP and may permit identification of different protein interactions compared to NP-40/TritonX-100. CHAPS is a zwitterionic detergent that is harsher than the nonionic detergents and very effective at solubilizing membranes. Some protein

interactions are preserved with CHAPS, as it is less harsh than ionic detergents like SDS or sodium deoxycholate (DOC). Only extremely stable protein interactions will be preserved using RIPA buffer containing 0.1% SDS, 1% DOC, and 1% NP-40. Ionic detergents are not recommended for IP or co-IP. When optimizing the IP, start by using NP-40, Triton X-100, or Digitonin at 1% and increase stringency if the bait protein is not recovered from membranes. Increase the stringency by increasing the nonionic detergent concentration to 2% or use CHAPS at 1–2%. If very few proteins are found to interact with the bait protein, decrease the stringency of the solubilization to ensure you can detect some reported protein interactions. Pay close attention to the temperature of the reactions at all times to maintain 0–4 °C, as the critical micelle concentration (CMC) of nonionic detergents is variable across temperature ranges, and slight fluctuations will alter the reproducibility of results [17].

28. Immunoblot analysis should be performed on both WT and KO fractions to ensure the gradients were similarly harvested.

29. Integral membrane proteins are preferred as fractionation markers; however, organelle activity and integrity is preserved during this isosmotic procedure, allowing for the identification and enzyme assay of soluble, organelle matrix proteins.

30. Buffer replacement is required, as high concentrations of OP may impede the diffusion of larger protein complexes and the formation of antibody–antigen interactions.

31. At this point the protein concentration of the lysates should be equal between the 2 genotypes (volume between 1 and 12.5 mL).

32. The required mass of IP material must be determined experimentally; however, ~5 mg of starting material (enriched gap junction fraction lysate) is a recommended minimum.

33. This step removes any protein from the lysate that nonspecifically binds to the bead matrix.

34. Alternate bead washing techniques. Depending on the stringency of the detergent, the nature of the tissue, proteins may nonspecifically bind to the beads and to the bait protein. If there are many nonspecific proteins eluting in the KO purification or in the control isotype purification, wash the beads more, or reevaluate the detergent choice. To modify the washes, try adding a detergent wash and further PBS washes. Do not use a detergent more stringent than used during the IP. The salt concentration in the wash buffer may also be titrated to a maximum stringency of 1000 mM. Increasing the bead wash stringency can reduce false-positive interactions but may prevent the identification of transient or weak protein interactions.

35. If the antibodies were not coupled to the bead matrix using DMP and there is no centrifugal vacuum concentrator available, proteins may be eluted from the beads using 1–2 bead volumes 2× SDS sample buffer containing 10% BME. Boil the beads in sample buffer for 5 min and centrifuge for 1 min at $1000 \times g$. Reserve the bead supernatant.

36. Reserve 5–10% of resuspended IP product for verification of IP success by immunoblotting.

37. Anything that touches the gel at this point must be wiped clean with 70% ethanol and care must be taken to ensure that no keratin or dust is introduced to the gel environment. Limit handling of the gel if at all possible, even with gloves.

38. Wipe the blade with 70% ethanol before use and change the blade between each band.

39. It is crucial that mass spectrometry identification of protein interactions from replicate IPs be confirmed biochemically to limit false-positive reports. Prioritize the biochemical validation of protein identities assigned from 1 or more unique peptides, and identities assigned from peptides that cover the largest % of the protein sequence, with the highest Mascot ion scores (probability that the MS/MS spectrum matches to the stated peptide). Mascot scores depend on the type and quality of the dataset. However, mean scores representing correct protein identities are usually around ~100 [18]. For large datasets, application of a 1% false-discovery rate (FDR) calculation should be applied to limit false-positive protein identity assignments [19, 20]. As proof that the experiment is able to replicate the literature, prioritize the validation of previously identified interacting proteins, in addition to validating novel proteins identified in the screen.

40. A positive reciprocal co-IP does not prove that proteins directly interact, but is a strong piece of evidence that the proteins are contained within the same complex. A validated reciprocal co-IP occurs if an IP for bait protein identifies interacting protein in the lysate and IP for interacting protein identifies bait protein in the lysate (by immunoblotting). Three main reactions should be performed: (1) Positive reaction: agarose beads coupled to primary antibody of choice with WT lysate applied; (2) negative reaction: agarose beads coupled to primary antibody of choice with KO lysate applied; and (3) control reaction: agarose beads coupled to an isotype-matched IgG antibody with WT lysate applied (e.g., mouse IgG antibody added to the agarose beads at the same concentration/bead ratio as the anti-mouse primary Cx32 antibody).

41. Scale each reaction down to 25–50 µL of beads and 250 µg of total protein. Maintain the same antibody:bead ratio as optimized in the original experiment.

42. The same floated membrane fractions prepared for the MS/MS experiment may also be used for the validation co-IP experiment.

Acknowledgements

This work was supported by the Canadian Institutes of Health Research (CIHR) AWT143081 and Canadian Institutes of Health Research (CIHR) Training Program in Neurodegenerative Lipidomics (CTPNL) (TGF-96121 to SALB). SF received scholarships from CTPNL, CIHR (Banting and Best Ph.D. Scholarship), Ontario Graduate Scholarship (OGS) Queen Elizabeth II Graduate Scholarship in Science and Technology.

References

1. Decrock E, De Bock M, Wang N et al (2015) Connexin and pannexin signaling pathways, an architectural blueprint for CNS physiology and pathology? Cell Mol Life Sci 72:2823–2851

2. Herve JC, Derangeon M, Sarrouilhe D et al (2012) Gap junctional channels are parts of multiprotein complexes. Biochim Biophys Acta 1818:1844–1865

3. Zhou JZ, Jiang JX (2014) Gap junction and hemichannel-independent actions of connexins on cell and tissue functions--an update. FEBS Lett 588:1186–1192

4. Dbouk HA, Mroue RM, El-Sabban ME et al (2009) Connexins: a myriad of functions extending beyond assembly of gap junction channels. Cell Commun Signal. doi:10.1186/1478-811X-7-4

5. Herve JC, Bourmeyster N, Sarrouilhe D (2004) Diversity in protein-protein interactions of connexins: emerging roles. Biochim Biophys Acta 1662:22–41

6. Giepmans BN (2004) Gap junctions and connexin-interacting proteins. Cardiovasc Res 62:233–245

7. Fowler SL, Akins M, Zhou H et al (2013) The liver connexin32 interactome is a novel plasma membrane-mitochondrial signaling nexus. J Proteome Res 12:2597–2610

8. Kanter HL, Laing JG, Beyer EC et al (1993) Multiple connexins colocalize in canine ventricular myocyte gap junctions. Circ Res 73:344–350

9. Altevogt BM, Paul DL (2004) Four classes of intercellular channels between glial cells in the CNS. J Neurosci 24:4313–4323

10. Berggard T, Linse S, James P (2007) Methods for the detection and analysis of protein-protein interactions. Proteomics 7:2833–2842

11. Lalonde S, Ehrhardt DW, Loque D et al (2008) Molecular and cellular approaches for the detection of protein-protein interactions: latest techniques and current limitations. Plant J 53:610–635

12. Jarvik JW, Telmer CA (1998) Epitope tagging. Annu Rev Genet 32:601–618

13. Laird DW, Jordan K, Thomas T et al (2001) Comparative analysis and application of fluorescent protein-tagged connexins. Microsc Res Tech 52:263–272

14. Lagny TJ, Bassereau P (2015) Bioinspired membrane-based systems for a physical approach of cell organization and dynamics: usefulness and limitations. Interface Focus. doi:10.1098/rsfs.2015.0038

15. Wieckowski MR, Giorgi C, Lebiedzinska M et al (2009) Isolation of mitochondria-associated membranes and mitochondria from animal tissues and cells. Nat Protoc 4:1582–1590

16. Cox B, Emili A (2006) Tissue subcellular fractionation and protein extraction for use in mass-spectrometry-based proteomics. Nat Protoc 1:1872–1878

17. Chen L-J, Lin S-Y, Huang C-C et al (1998) Temperature dependence of critical micelle concentration of polyoxyethylenated nonionic surfactants. Colloids Surf B Physiochem Eng Aspects 135:175–181

18. Stead DA, Preece A, Brown AJ (2006) Universal metrics for quality assessment of protein identifications by mass spectrometry. Mol Cell Proteomics 5:1205–1211

19. Elias JE, Haas W, Faherty BK et al (2005) Comparative evaluation of mass spectrometry platforms used in large-scale proteomics investigations. Nat Methods 2: 667–675

20. Wang G, Wu WW, Zhang Z et al (2009) Decoy methods for assessing false positives and false discovery rates in shotgun proteomics. Anal Chem 81:146–159

Chapter 9

Scrape Loading/Dye Transfer Assay

Pavel Babica, Iva Sovadinová, and Brad L. Upham

Abstract

The scrape loading/dye transfer (SL/DT) technique is a simple functional assay for the simultaneous assessment of gap junctional intercellular communication (GJIC) in a large population of cells. The equipment needs are minimal and are typically met in standard cell biology labs, and SL/DT is the simplest and quickest of all the assays that measure GJIC. This assay has also been adapted for in vivo studies. The SL/DT assay is also conducive to a high-throughput setup with automated fluorescence microscopy imaging and analysis to elucidate more samples in shorter time, and hence can serve a broad range of in vitro pharmacological and toxicological needs.

Key words Dye coupling, Dye transfer, Ex vivo assessment, Gap junctional intercellular communication assessment, High throughput, In vitro assay, Incision loading, Lucifer Yellow, Scalpel loading, Scrape loading, Tracers

1 Introduction

Dye coupling methods are by far the most frequently used assay for the assessment of GJIC, mainly because of their ease of use. Of all the techniques used to measure GJIC, the scrape loading/dye transfer (SL/DT) assay is the fastest and simplest. Most protocols are modification of the one first reported by El-Fouly et al. [1]. This technique has since been widely used to elucidate the GJIC status of many cell types in various biological circumstances in different scientific areas such as carcinogenesis, embryogenesis, growth control, or endocrine disruption (for review, see [2–6]). This visual method allows to assess GJIC in a large population of cells. It is therefore particularly useful when a large screen of multiple conditions is required or when different regions of a cell monolayer have to be compared within the same culture dish [2]. The SL/DT assay can be effectively used as a tool to determine the qualitative and quantitative presence or absence of GJIC as well as demonstrate the concentration-dependent inhibition of GJIC [3, 7–9].

Mathieu Vinken and Scott R. Johnstone (eds.), *Gap Junction Protocols*, Methods in Molecular Biology, vol. 1437, DOI 10.1007/978-1-4939-3664-9_9, © Springer Science+Business Media New York 2016

The SL/DT assay relies on the introduction of small (MW <900), nonpermeable dyes (for review, *see* [2, 10]) into living cells that are traced in their intercellular movement through gap junctions. As the reference dye, dilithium salt of Lucifer Yellow hydrazine (LY, MW 457, negatively charged) is the most popular dye currently in use. This tracer has a high fluorescence efficiency, which ensures its detection in minute levels [10]. LY is introduced by scraping a monolayer of cells and becomes incorporated by cells along the scrape, presumably as a result of some mechanical perturbation of the membrane (Fig. 1). As normal permeability is reestablished, the LY becomes trapped within the cytoplasm and move from the dye-loaded cells into adjacent ones connected by functional gap junctional channels [2]. This dye transfer is monitored and quantified by fluorescent microscopy in multiple cells almost simultaneously. The amount of dye transferred from one cell to its neighbor that it is in contact with is dependent on the number of gap

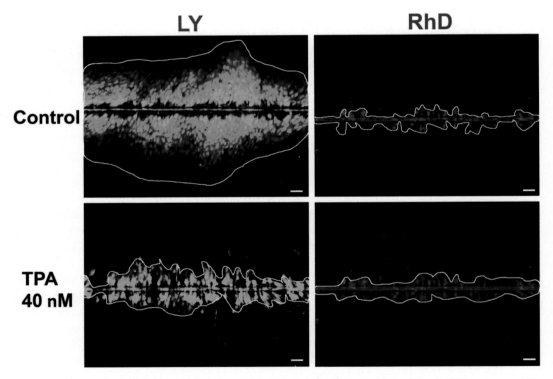

Fig. 1 *Scrape load dye transfer analysis in mouse Sertoli cells.* Images obtained by SL/DT assay, applying Lucifer Yellow CH dilithium salt (LY, MW 457), which transfers through functional gap junction channels, and rhodamine-dextran (RhD, MW 10,000), which is retained in the scraped cells. The GJIC function is evaluated by analyzing net transfer of LY (the area at which LY diffuses), excluding RhD-stained regions. GJIC after the 1-h exposure of mouse Sertoli TM4 cells to the model tumor promoter, TPA (12-O-tetradecanoylphorbol-13-acetate), at the concentration of 40 nM was reduced to FOC (the fraction of the control) = 0.13 when compared with the solvent control. Scale bar = 50 μm

junctions that are coupled and the gating properties of individual channels. The distance or area at which the dye diffuses during a certain period away from the scrape line is a quantitative measurement of GJIC capacity. To determine which cells are initially loaded after the scrape, other fluorescent dyes (e.g., rhodamine-dextran, MW 10,000, or dialkylcarbocyanine) that are too large to traverse the gap junction channel are concurrently used with the diffusional dye to serve as an additional control (Fig. 1). These large fluorescent macromolecules that cannot diffuse across gap junctions are useful in ensuring that the intercellular transfer of the gap junction diffusible-reference dye is actually dependent on gap junctions and is not accounted for by alternative pathways, such as cytoplasmic membrane fusions, cytoplasmic bridge formation at the end of mitosis or due to membrane damage, which can occur after scraping [1, 11].

The SL/DT assay is an invasive technique but has been successfully demonstrated to assess compounds that disrupt GJIC. However, this assay is not conducive in studying GJIC in small cell populations, particularly between cell pairs, and also in cultures with low cell densities, or when the extent of junctional coupling is small, or when specific cells need to be observed [2]. In addition, the GJIC status of cell types of irregular shape is not easily quantified using this assay. For example, GJIC of neuronal cells or long spindly fibroblast cannot be easily quantified because the distance or area at which the dye diffuses cannot be easily "trackable" and quantified [11]. This approach is also not well suited to three-dimensional (3D) systems. The local activation of molecular fluorescent probe (LAMP) method has been recently improved (the so-called infrared-LAMP assay) and allows to examine cell–cell coupling in three dimensions [12, 13]. However, for the time being, two-dimensional cell culture systems still serve as a valuable tool in cell biology and toxicology research [5].

The major advantages of the SL/DT are as follows: (1) simplicity, (2) not a necessity of the special equipment or skills that are needed for other methods such as microinjection, (3) a rapid and simultaneous assessment of GJIC in a large number of cells, (4) conducive to a high-throughput setup with automated fluorescence microscopy imaging and analysis, and (5) its adaptation for in vivo studies followed by ex vivo assessment of GJIC in tissue slices from experimental animals [5]. An ex vivo GJIC assay, the incision loading/dye transfer method (IL/DT), is very similar to the in vitro protocol [14–16]. The IL/DT may be useful for rapidly screening tumorigenic compounds for setting doses for studies of carcinogenesis [14].

The basic technique described in this chapter has been adapted after the method of El-Fouly [1]. Rather than more invasive scrape with rubber policeman or wooden probe, the dye loading step in this protocol involves a clean cut with a sharp blade, such as surgical scalpel. This modified technique can be thus called scalpel loading/dye transfer assay and is amendable to many cell types with minimal or no modifications. This assay has been extensively applied to determine changes in GJIC in a wide variety of mammalian (including human) cell types after treatment with many kind of toxicants such as tumor promoters, endocrine disruptors, pesticides, or developmental toxicants. Additionally, this assay has been successful in screening for compounds that can either prevent toxicant-induced disruption of GJIC or reverse the effects of these toxicants or endogenous oncogenes.

2 Materials

1. Cells of interest and appropriate media.

2. Cell culture plates, e.g., 35 mm dishes, or multiwell plates, e.g., 6/12/24/48/96 wells.

3. Phosphate buffered saline (PBS) buffer with calcium and magnesium (CaMg-PBS; *see* **Note 1**): 137 mM NaCl (8 g/L), 2.68 mM KCl (0.2 g/L), 8.10 mM Na_2HPO_4 (1.15 g/L), 1.47 mM KH_2PO_4 (0.2 g/L), 0.68 mM $CaCl_2$ (0.075 g/L), 0.49 mM $MgCl_2$ (0.047 g/L). The pH is adjusted to 7.2. The CaMg-PBS is filter sterilized and can be stored at room temperature.

4. LY-dye solution: 1 mg/mL Lucifer Yellow CH dilithium salt (LY, MW 457) (*see* **Note 2**) and 1 mg/mL rhodamine-dextran (RhD, MW 10,000, optional, *see* **Note 3**) in CaMg-PBS. The solution is filter sterilized and can be stored for weeks in the dark at 4 °C (*see* **Note 4**).

5. Surgical scalpel blade or micro-knife with a curved, flat, or needle blade as appropriate (Fig. 2) to fit into the cell culture plasticware used (*see* **Note 5**).

6. 10 % Formalin solution (i.e., approximately 4 % formaldehyde) in CaMg-PBS (*see* **Note 6**).

7. 25 mL pipettes with a pipette aid or manual bulb, automatic pipette 100–1000 µL or Pasteur pipette with a bulb, waste container, parafilm, aluminum foil.

8. Inverted epifluorescent microscope or confocal microscope with appropriate filters (LY: excitation at 428 nm, emission at 536 nm; RhD: excitation at 555 nm, emission at 580 nm).

9. Camera, CCD or CMOS camera coupled to the microscope, computer, image acquisition and analysis software.

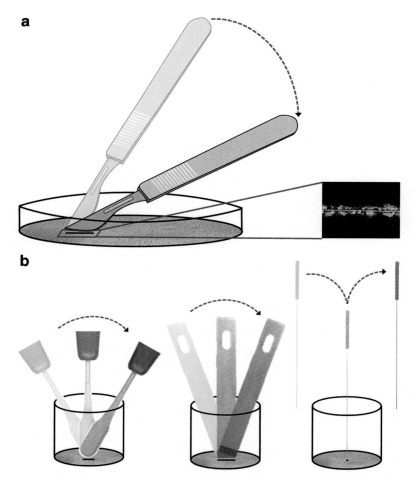

Fig. 2 *The scrape loading procedure.* (**a**) Scrape loading technique done by surgical scalpel with a curved blade in 35 mm dishes. (**b**) Different types of blades suitable for SL/DT assay in microplate wells, including a curved blade (*left*), a flat blade (*middle*), and an acupuncture needle (*right*)

3   Methods

3.1   General Procedure

1. Grow cells to confluency in suitable cultivation medium under appropriate conditions for the desired cell type (*see* **Note 7**).

2. Visually check the health of the cells for expected morphologies and sterile conditions prior to the start of the experiments.

3. Remove cells from the incubator and discard medium by gently pouring off the medium or by siphoning with a pipette into a waste container (*see* **Notes 8** and **9**).

4. Rinse cells gently three times with CaMg-PBS using a pipette (*see* **Note 10**) and discard the CaMg-PBS by gently pipetting off into a waste container.

5. Add sufficient LY-dye solution (warmed up to 37 °C) to cover the cell monolayer (*see* **Note 11**).

6. Load the dye into the cells by gently placing the tip of a surgical steel blade with a curved edge in contact with the cell monolayer and then rolling the blade in one direction over its curved edge as indicated in Fig. 2a (*see* **Note 12**). If using multiwell plates, use a micro-knife with a curved or flat blade, or an acupuncture needle (*see* **Note 5**) to gently prick the cells (Fig. 2b).

7. Typically, three cuts are done for each dish or multiwell plate well. The areas for cell loading are randomly selected in the central part of the plate/well (*see* **Note 13**).

8. Incubate the dish or multiwell plate, undisturbed and under minimum illumination for 3–6 min at room temperature to allow the LY dye to travel through several adjacent cell layers (3 and more) via functional gap junctions (*see* **Note 14**).

9. Cover the plate, to limit the exposure of the LY-dye solution to light during the incubation to avoid fluorescence photobleaching of the dye.

10. Aspirate the LY-dye solution from plate/wells (*see* **Note 15**) and then rinse cells three times with CaMg-PBS to remove all extracellular dye (*see* **Notes 10** and **16**).

11. Fix the cells by adding sufficient 10% formalin solution to cover the cells (*see* **Notes 11** and **17**).

12. The cells can be viewed immediately using an inverted epifluorescence microscope or confocal microscope with appropriate filters.

13. Acquire three representative LY/RhD images per plate/well (*see* **Note 18**).

14. A bright field or phase contrast image should be acquired for each field of view (*see* **Note 19**).

15. Fixed cells can be air-dried overnight, stored in the dark for extended periods (months with no detectable decrease in dye intensity), and rehydrated with formalin solution for viewing.

16. Store the plates sealed in parafilm and covered in aluminum foil.

3.2 Quantification of GJIC Using Morphometric Software

1. The degree of GJIC can be measured using a variety of methods (*see* **Note 20**) [17].

2. The fluorescent distance or the area of the dye spread can be quantified using a morphometric software package. We use ImageJ, a free public domain imaging software package from

the National Institute of Health (http://imagej.nih.gov/ij/), with a subroutine for determining fluorescence area of a fluorescent image.

3. To obtain a corrected fluorescence area value for LY and RhD, the fluorescent areas of digitized images (e.g., Fig. 1) are subtracted from background fluorescence obtained at an area of the monolayer well away from any scrape lines within the same test plate (*see* **Note 21**).

4. To calculate the net transfer of LY, the areas of LY and RhD fluorescence are subtracted from each other for each field of view (i.e., Area^{LY}-Area^{RhD}) (*see* **Note 22**).

5. The net LY areas of individual images can then be normalized to the averaged net area from negative or solvent control dishes to obtain the fraction of the control (GJIC-FOC)

$$\text{GJIC FOC}_{\text{Treatment}} = \frac{\left(\text{Area}_{\text{Treatment}}^{LY} - \text{Area}_{\text{Treatment}}^{RhD}\right)}{\left(\text{Area}_{\text{Control(Averaged)}}^{LY} - \text{Area}_{\text{Control(Averaged)}}^{RhD}\right)}$$

6. GJIC-FOC values of individual images can be then grouped by the individual dishes or treatment conditions for further data evaluation and statistical analyses to allow relative comparisons between the control and the treatments, for concentration and time response analyses, or for comparisons between independent experiments.

4   Notes

1. Calcium chloride must be added before the magnesium chloride to avoid irreversible precipitation of the salts. This buffered solution should not be sterilized by autoclave because of salt precipitation but should be rather filter sterilized through a sterile 0.22 μm filter. Calcium and magnesium cations are added to the PBS buffer to maintain cell adhesion and prevent a monolayer of cells from lifting, i.e., detaching from the bottom of culture dishes or plates. However, some cell types may be sensitive to this level of Ca^{2+}. This problem may be alleviated by preparing the LY-dye solution in CaMg-free PBS.

2. Some junction channels exclude anionic molecules like LY, for example, connexin 45 [2, 18, 19]. For these channels, smaller cationic dyes such as biotin conjugate (e.g., Neurobiotin [16, 20] or Biocytin [21]) are recommended. To be detected, biotin conjugates should be visualized with either streptavidin coupled to a fluorochrome (cyanine dyes, fluorescein, or rhodamine) or to horseradish peroxidase. If the connexin channels of a particular cell type are unknown, then SL/DT using LY as the transfer dye is not enough to elucidate GJIC status in these cells.

3. Use LY-dye solution with RhD if there is a need to identify which cells are initially loaded with the dye. RhD is a large dye that does not pass through gap junctions, while LY does pass through gap junctions.

4. The LY-dye solution must be warmed to 37 °C before use on the cells.

5. Surgical scalpel with a curved blade is suitable for 35 mm dishes (Fig. 2a). Micro-knives and blade holders with curved blades, flat blades, or ultrafine needle blades (e.g., acupuncture needles) with dimensions fitting multiwell plate wells are suitable for SL/DT in setups allowing for higher throughput (Fig. 2b).

6. The formalin solution should be prepared in a chemical fume hood and safety goggles and gloves should be worn. Shelf-life of this solution is approximately 3 months.

7. The growth phase at which a SL/DT experiment is done is critical. Typically, cells which have reached confluency and are no longer actively dividing ("contact inhibition") are the most suitable for the SL/DT assay. For new cells that have not been previously assessed for GJIC, the growth conditions, i.e., cell seeding density as well as culture time, must be optimized. In addition, the passage number must be noted during GJIC experiments because the passage number can play a significant role in functioning GJIC in a given cell type. Many cells have abundant PDGF (platelet-derived growth factor) receptors. PDGF inhibits GJIC in several cell lines [22], thus conducting GJIC experiments with cells grown in medium supplemented with fetal bovine serum (FBS) poses problem due to the high levels PDGF in FBS. Transferring the cells to FBS free for 2–4 h can overcome this problem. Due to the unnatural two-dimensional environments of the traditional in vitro assays, some cell types may not establish GJIC in the traditional medium or plastic. Some cell types need specific culture conditions to express functional GJIC such as extracellular matrix-coated plates (e.g., mammary CID-9 cell line [23]) or low or high calcium medium (e.g., mouse epidermal cell line [24]). Some cell types such as mouse testicular cell lines TM3 and TM4 can detach from the bottom of tissue culture plates during the washing steps. Growing cells on gelatin-coated plates can overcome this problem.

8. Culture medium containing hazardous waste must be properly disposed.

9. This and the subsequent steps are usually done on the lab bench at room temperature.

10. We typically use a 25 mL pipette, which is sufficient volume to rinse several 35 mm cell culture dishes or a 6- to 96-well multiwell plate.

11. We typically use about 1 mL of solution per 35 mm cell culture dish or per well of 6- or 12-well plate, 0.5 mL per well of 24-well plate, 0.25 mL per well of 48-well plate, 0.1 mL per well of 96-well plate.

12. The key principle of proper loading technique is to put the tip or apex of the blade to the bottom of the dish or multiwell plate well and then roll the blade over its cutting edge against the cell monolayer or to gently prick the monolayer with the tip of a thin acupuncture needle. This action is minimally invasive and provides a very clean line or spot of loading. Do not slice or scratch the monolayer, but only apply gentle pressure to minimize physical effects of this step. A sharp blade or point is important to prevent a large separation or empty hole between the cells that were loaded resulting in high variability.

13. This scrape loading step is done at the lab bench without use of a microscope. If you want to assess GJIC in any specific area of special interests that you found during microscopic examination, then use a marker to indicate where the scrape line needs to be placed, or use the microscope. The three cuts per dish or well should be aligned in parallel, and in the case of multiwell plates also geometrically parallel with the base of the well.

14. The optimal incubation time varies between different cell types, depending on the level of communication and attachment properties of the cells [8]. Incubation up to 10 min might be required for some cell cultures. For a set experiment duration, different rates of dye diffusion through homotypic channels is correlated to the number of gap junction channels [25]. If processing several dishes or multiwell plates in parallel, work in a timely manner to make sure that the incubation time after the dye loading step will be the same for all dishes or wells.

15. LY-dye solution can be collected and reused. We reuse the dye solution for approximately 10 experiments, when stored in dark and refrigerated. We filter the solution through 0.22 μm syringe filter, if needed.

16. The washing step is very important step because of the reduction of background fluorescence. Even in the absence of a scrape line, some LY can be incorporated into the cells as evidenced by nonspecific background fluorescence. It also binds to cell components after aldehyde fixation [10].

17. The fixation of cells is optional and can be skipped. The cells in CaMg-PBS or medium can also be observed without fixation, but the dye will continue to travel through the cell layers and become overly diffuse to observe.

18. The cells from experimental conditions where the highest and lowest level of GJIC are expected (e.g., negative and positive controls) should be used first to adjust or check the microscope and image acquisition settings. The plates should be positioned so the line or spot of dye-loaded cells will be in the center of the microscope field of vision, and the line also parallel with the horizontal line of image field. Camera exposure time and other image acquisition settings (e.g., excitation source intensity or fluorescence attenuator, camera binning, image brightness, contrast and gamma correction) should be adjusted in a way that LY- or RhD-stained cells can be clearly discriminated from the background, i.e., from the cells whose fluorescence intensity is comparable to all the other cells in regions distant from the cut. However, the background cells should not turn out completely black in the image (too dim images), since such condition might lead to underestimation of GJIC. The used combination of the objective magnification (typically 5–20× objective), digital camera (e.g., C-Mount adapter magnification, size of the imaging sensor), and other image acquisition settings should allow to fit within one image (one field of view) not only all LY-stained cells in the direction perpendicular to the cut, but also part of the background, so to assure most accurate quantification GJIC and to prevent underestimation of communication in the cells from the experimental conditions with the most intense GJIC.

19. Bright field or phase contrast images from the same field of view as LY/RhD fluorescence images can offer additional visual information on the cells, such as the effects of experimental treatments on cell morphology, growth, confluency, and attachment, and also provide additional information to discriminate between the reduction of LY-stained area due to inhibition of GJIC, as compared to reduced dye transfer as a function of subconfluency or cell detachment issues.

20. An alternative method of quantitating GJIC is by counting the rows of fluorescent cells from the scrape line. This method is useful and more appropriate when comparing populations of different cell type, size, and growth state [26].

21. The most frequent problem is the high intensity of background fluorescence, so the cells stained by LY due to the dye transfer cannot be discriminated from the cells in the background. The most common contributions to background fluorescence are as follows: (1) an insufficient rinse of the extracellular dye, (2) treating cells with cytotoxic concentrations of the chemical (an uptake of dye by cells that

were not scraped or near to scrape line indicating disrupted cell membranes or detachment of the cells from the plate during the rinse step), and (3) "overconfluent cells," i.e., cells have been confluent for more than one day and have started to produce significant extracellular matrix (LY binds to extracellular matrix). Dim images are almost always problem related to the photobleaching of LY solution, especially if it is being recycled and reused. In the short term, this issue can be compensated for by increasing exposure time for fluorescence image acquisition. Fresh LY solutions will usually alleviate this problem.

22. A more simplified version of the SL/DT assay is to load cells with only LY and not the RhD. The results from these experiments typically give FOC values very similar to those that include RhD. However, when measuring the dye fronts, the loaded cells cannot be identified, and thus cannot be subtracted from the calculations resulting in values that are always above zero. This issue can be circumvented by introducing a positive control into the experimental design, i.e., treatment with a known inhibitor with GJIC (such as 12-O-tetradecanoyl-13-phorphol acetate), which will induce complete inhibition of GJIC. The net LY dye transfer can be then calculated by subtracting the average area of LY-stained cells in the positive control from LY-stained areas of all images. Adjusted areas of the experimental treatments can be then compared to the averaged adjusted area of the negative or solvent control:

$$
\text{GJIC FOC}_{\text{Treatment}} = \frac{\left(\text{Area}_{\text{Treatment}}^{\text{LY}} - \text{Area}_{\text{Positive control (Averaged)}}^{\text{LY}} \right)}{\left(\text{Area}_{\text{Negative control (Averaged)}}^{\text{LY}} - \text{Area}_{\text{Positive control (Averaged)}}^{\text{LY}} \right)}
$$

This approach is suitable only for in vitro models with well characterized GJIC, where complete inhibition of GJIC can be reproducibly induced and reliably recognized. However, this greatly simplifies the assay, particularly at the microscopy step where only one dye needs to be assessed.

Acknowledgements

This work was financially supported by the National Sustainability Programme of the Czech Ministry of Education, Youth and Sports (LO1214) and the RECETOX research infrastructure (LM2015051), and NIEHS grant #R01 ES013268-01A2 to Upham.

References

1. Elfouly MH, Trosko JE, Chang CC (1987) Scrape-loading and dye transfer—a rapid and simple technique to study gap junctional intercellular communication. Exp Cell Res 168:422–430

2. Abbaci M, Barberi-Heyob M, Blondel W et al (2008) Advantages and limitations of commonly used methods to assay the molecular permeability of gap junctional intercellular communication. Biotechniques 45:33–58

3. Klaunig JE, Shi Y (2009) Assessment of gap junctional intercellular communication. Curr Protoc Toxicol Chapter 2, Unit2 17

4. Trosko JE, Ruch RJ (2002) Gap junctions as targets for cancer chemoprevention and chemotherapy. Curr Drug Targets 3:465–482

5. Upham BL (2011) Role of integrative signaling through gap junctions in toxicology. In: Maines MD (ed) Curr Protoc Toxicol, vol Chapter 2. p Unit 2.18

6. Vinken M, Doktorova T, Decrock E et al (2009) Gap junctional intercellular communication as a target for liver toxicity and carcinogenicity. Crit Rev Biochem Mol Biol 44:201–222

7. Loch-Caruso R, Caldwell V, Cimini M et al (1990) Comparison of assays for gap junctional communication using human embryocarcinoma cells exposed to dieldrin. Fundam Appl Toxicol 15:63–74

8. Opsahl H, Rivedal E (2000) Quantitative determination of gap junction intercellular communication by scrape loading and image analysis. Cell Commun Adhes 7:367–375

9. Sovadinova I, Babica P, Boke H et al (2015) Phosphatidylcholine specific PLC-induced dysregulation of gap junctions, a robust cellular response to environmental toxicants, and prevention by resveratrol in a rat liver cell model. PLoS One 10:e0124454

10. Meda P (2000) Probing the function of connexin channels in primary tissues. Methods 20:232–244

11. Trosko JE, Chang CC, Wilson MR et al (2000) Gap junctions and the regulation of cellular functions of stem cells during development and differentiation. Methods 20:245–264

12. Dakin K, Zhao Y, Li WH (2005) LAMP, a new imaging assay of gap junctional communication unveils that Ca2+ influx inhibits cell coupling. Nat Methods 2:55–62

13. Yang S, Li WH (2009) Assaying dynamic cell-cell junctional communication using noninvasive and quantitative fluorescence imaging techniques: LAMP and infrared-LAMP. Nat Protoc 4:94–101

14. Sai K, Kanno J, Hasegawa R et al (2000) Prevention of the down-regulation of gap junctional intercellular communication by green tea in the liver of mice fed pentachlorophenol. Carcinogenesis 21:1671–1676

15. Upham BL, Park JS, Babica P et al (2009) Structure-activity-dependent regulation of cell communication by perfluorinated fatty acids using in vivo and in vitro model systems. Environ Health Perspect 117:545–551

16. Goliger JA, Paul DL (1995) Wounding alters epidermal connexin expression and gap junction-mediated intercellular communication. Mol Biol Cell 6:1491–1501

17. McKarns SC, Doolittle DJ (1992) Limitations of the scrape-loading dye transfer technique to quantify inhibition of gap junctional intercellular communication. Cell Biol Toxicol 8:89–103

18. Pastor A, Kremer M, Möller T et al (1998) Dye coupling between spinal cord oligodendrocytes: differences in coupling efficiency between gray and white matter. Glia 24:108–120

19. Risley MS, Tan IP, Farrell J (2002) Gap junctions with varied permeability properties establish cell-type specific communication pathways in the rat seminiferous epithelium. Biol Reprod 67:945–952

20. Bittman K, Becker DL, Cicirata F et al (2002) Connexin expression in homotypic and heterotypic cell coupling in the developing cerebral cortex. J Comp Neurol 443:201–212

21. Rouach N, Segal M, Koulakoff A et al (2003) Carbenoxolone blockade of neuronal network activity in culture is not mediated by an action on gap junctions. J Physiol 553:729–745

22. Yamasaki H, Naus CC (1996) Role of connexin genes in growth control. Carcinogenesis 17:1199–1213

23. El-Sabban ME, Sfeir AJ, Daher MH et al (2003) ECM-induced gap junctional communication enhances mammary epithelial cell differentiation. J Cell Sci 116:3531–3541

24. Jongen WM, Fitzgerald DJ, Asamoto M et al (1991) Regulation of connexin 43-mediated gap junctional intercellular communication by Ca2+ in mouse epidermal cells is controlled by E-cadherin. J Cell Biol 114:545–555

25. Czyz J, Irmer U, Schulz G et al (2000) Gap-junctional coupling measured by flow cytometry. Exp Cell Res 255:40–46

26. Vaz-de-Lima BB, Ionta M, Machado-Santelli GA (2008) Changes in cell morphology affect the quantification of intercellular communication. Micron 39:631–634

Chapter 10

Microinjection Technique for Assessment of Gap Junction Function

Michael D. Fridman, Jun Liu, Yu Sun, and Robert M. Hamilton

Abstract

Gap junctions are essential for the proper function of many native mammalian tissues including neurons, cardiomyocytes, embryonic tissues, and muscle. Assessing these channels is therefore fundamental to understanding disease pathophysiology, developing therapies for a multitude of acquired and genetic conditions, and providing novel approaches to drug delivery and cellular communication. Microinjection is a robust, albeit difficult, technique, which provides considerable information that is superior to many of the simpler techniques due to its ability to isolate cells, quantify kinetics, and allow cross-comparison of multiple cell lines. Despite its user-dependent nature, the strengths of the technique are considerable and with the advent of new, automation technologies may improve further. This text describes the basic technique of microinjection and briefly discusses modern automation advances that can improve the success rates of this technique.

Key words Microinjection, Gap junction, Cellular communication, Connexin, Dye transfer

1 Introduction

For many mammalian tissues, there is a reliance on the ability for cells to communicate with each another. Cardiac tissue, as a prime example, relies on the coordination of cellular contraction such that all cells are recruited to be able to work as a single unit—a unit described as a syncytium. To facilitate this process, membrane channels, known as gap junctions, link cells together creating a pore that allows the diffusion of hydrophilic, size-limited (i.e., <1 kDa) particles including ions, nutrients, metabolites, and secondary messengers from the cytoplasm of one cell to the next [1]. These gap junctions are composed of 2 half-channels called connexons [2, 3], which are each composed of 6 connexin subunits. Mutations in these connexin proteins result in a wide variety of disease states including hearing loss, cardiac arrhythmia, and other complex syndromes [4–6]. Methods of quantifying gap junction function are therefore instrumental in developing an understanding of these conditions and how to address them.

Mathieu Vinken and Scott R. Johnstone (eds.), *Gap Junction Protocols*, Methods in Molecular Biology, vol. 1437,
DOI 10.1007/978-1-4939-3664-9_10, © Springer Science+Business Media New York 2016

There are myriad methods that can be utilized for gap junction assessment in the laboratory including the scrape/scratch method [7], electroporation [8], fluorescence redistribution after photobleaching [9], conductance measurement by dual-patch clamp [10], and microinjection [11, 12]. Each of these topics is discussed in greater detail in other chapters of this text. The scrape/scratch, electroporation, and microinjection techniques are similar in their reliance on introducing dye into cells and observing the dye distribution across communicating cells. The major differences between these techniques are the ease of learning, reproducibility of the results, and measurability of the result. For dynamic measurements, conductance-based measurements should be performed.

Microinjection, the process in which a small volume of a nontoxic tracer is injected into single cells, is a common method used to quantify gap junction function [13]. There are significant limitations, though, due to the technical skill required. Many studies have shown that interuser variability between those trained in the technique is significant with respect to cell viability, reproducibility, and length of training time [14]. Due to microinjection's challenging nature, the other dye-introduction techniques were developed, though the quality of their results compared to properly done microinjection is lower. Despite its complexity, microinjection continues to offer several distinct advantages including the ability to correlate morphological data with functional data from individually targeted cells, enabling time-lapse studies to quantify and visualize the kinetics of cell-to-cell communication, and the ability to directly compare gap junction communication in a variety of cell types [15]. With the microinjection technique, cardiomyocytes (e.g., HL-1 and iPSC-derived cardiomyocytes) have shown significantly higher gap junction communication than other cell lines such as HeLa and HEK293 cells. Furthermore, microinjection is unique in that nontracer compounds—for example, plasmids or proteins—can be injected into the cytoplasm or nucleus, although these topics are beyond the scope of this text.

Typically, experiments require a significant commitment of time, training hours, and resources to perform sufficient microinjection to gather reliable data. As such, a variety of software and hardware have been developed to facilitate micropipette location [16, 17], cell contact detection [18], and cell injection volume [19]. Recently, these components were combined into a single device capable of automated high-throughput microinjection, which eliminated much of the significant user dependence and time commitment to develop technique proficiency [20, 21]. Although automation will likely become an important technique due to its ability to facilitate a highly robust methodology, it is still in its infancy and the setup is complex. In this chapter, the general technique associated with the successful performance of the manual microinjection will be discussed.

2   Materials

2.1   Establishment of Cell Culture

1. Dulbecco's Modified Eagle Medium (DMEM) cell culture medium: DMEM with 10% fetal bovine serum (FBS) (*see* **Note 1**).

2. Claycomb cell culture medium: claycomb medium supplemented with 10% FBS, 100 µg/mL penicillin–streptomycin, 10 µM epinephrine, and 2 mM l-glutamine (*see* **Note 2**).

3. Incubator set to 37 ± 1 °C, humidity 95%, CO_2 5%.

4. Laminar airflow cabinet.

5. 35 mm fibronectin-coated plastic cell culture plates.

6. 37 °C water bath for warming solutions.

2.2   Microinjection

1. Inverted microscope with 200× magnification. A typical microinjection system is shown in Fig. 1 (*see* **Note 3**).

2. Micromanipulators for platform and micropipette holder.

3. Microinjectors.

4. Glass micropipette pulled via microfilament heating to achieve 1 mm outer and 0.5 mm inner diameter.

5. Microloaders to inject dye into micropipettes.

6. Micropipette puller.

7. Phosphate buffered saline (PBS), pH 7.4.

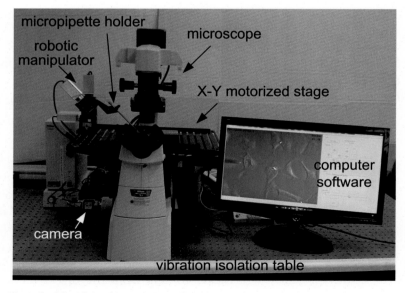

Fig. 1 Microinjection system for the measurement of gap junction communication

8. Dye solutions: the solutions used in these experiments are membrane-impermeable, water-soluble dyes. The low molecular weight dye was 8-hydroxypyrene-1,3,6-trisulfonic acid (HPTS, MW = 524.7 Da) which emits light in the green fluorescent range and is sufficiently small to pass through gap junctions. The high molecular weight dye was rhodamine-dextran (MW 10,000 Da) which emits light in the red fluorescent range and cannot pass through gap junctions due to its size. Dyes can be used as 2 mM solutions mixed at a 1:1 ratio within the micropipette (*see* **Note 4**).

2.3 Microscopy

1. Fluorescent microscope/confocal microscope.
2. Camera, computer, software.

3 Methods

3.1 Establishment of Cell Culture Monolayer

1. Use low-passage cells to begin your culture (*see* **Note 5**).
2. Plate cells on 35 or 60 mm dishes based on the number of cells you are hoping to inject. Seeding densities of 2×10^5 cells/plate are adequate for 35 mm dishes and 8×10^5 cells/plate for 60 mm dishes (*see* **Notes 6–8**). Place dish in 37 °C incubator and allow cells to grow to confluence, e.g., greater than 90 %.

3.2 Preparation of Cells for Microinjection

1. Warm all solutions prior to use (*see* **Note 9**).
2. Wash cells in 2 mL fresh culture medium to remove debris.
3. Wash in 1–2 mL PBS 3 times.
4. Place cells in 1–2 mL fresh culture medium and transport to microinjections station.

3.3 Preparation of the Microinjection Injection Apparatus

1. Place dish with culture medium (i.e., no cells) on the microinjection platform.
2. Dip the tip of the micropipette into dye solution (*see* **Note 10**).
3. Fill the micropipette with the remainder of the dye solution using a microinjector.
4. Gently tap or flick the micropipette to remove any air bubbles (*see* **Note 11**).
5. Mount the micropipette to the microinjection apparatus and micromanipulator. Angle the manipulator at 30–45° from the horizontal plane of the microscope stage.
6. Once micropipette is lowered into the culture medium, observe for dye release from the tip. There should be a consistent trickle coming from the tip. If there is not, adjust constant positive pressure, sometimes referred to as compensation pressure, until there is (*see* **Note 12**).

7. Set injection pressure to, at minimum, double the compensation pressure (*see* **Note 13**). This pressure will vary based on experiments performed and should be modified as needed based on cell size and type.

8. Set injection time based on cell line used and previously published information. For HEK, HeLa, and HL-1 cells, the injection time of 0.4 s was used with high success.

3.4 Microinjection of Cells

1. Take cells and place in injection chamber.

2. Focus microscope until the cytosol and membranes can be clearly seen (*see* **Notes 14** and **15**).

3. Dip tip of micropipette into dye solution (*see* **Note 10**).

4. Fill the micropipette with the remainder of the dye solution using a microinjector.

5. Gently tap or flick the micropipette to remove any air bubbles (*see* **Note 11**).

6. Mount the micropipette to the microinjection apparatus and micromanipulator. Angle the manipulator at 30–45° from the horizontal plane of the microscope stage.

7. Use "coarse" adjust to lower the micropipette into the dish and locate the tip by adjusting the X–Y plane of the micropipette until it comes into view (*see* **Note 16**).

8. Use "fine" adjustments to lower the micropipette into the appropriate visual plane over top of the cell of interest, as shown in Fig. 2a.

9. Slowly lower the micropipette tip until it is in contact with the membrane without puncturing (*see* **Note 17**).

10. Slowly and carefully lower the tip with "fine" adjustment until the micropipette visibly enters the cytosol.

11. Inject using the injection time/pressure optimized in prior experiments, as shown in Fig. 2b (*see* **Notes 18** and **19**).

12. Inject multiple cells in the monolayer as needed (*see* **Notes 20–23**).

3.5 Cell Analysis

1. Once sufficient cells are injected, cover the dish and wrap in foil to protect from further light exposure.

2. Cells should be washed twice with PBS to remove any dye that may be remaining in the culture medium.

3. Replace cells in 37 °C incubator for the appropriate time intervals decided prior to the experiment (*see* **Note 24**).

4. Place cells from incubator to fluorescent microscope to image using appropriate filters.

5. Measurement of the degree dye transfer can include the number of cells the dye has transferred to or the distance of dye travel.

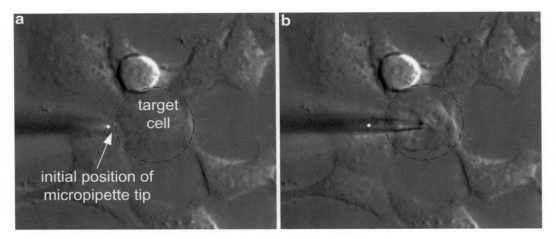

Fig. 2 *Microinjection of a single cell.* (**a**) Micropipette tip is positioned close to the target cell. (**b**) Fluorescent molecule injection into the cell after the tip has penetrated into the cell cytosol

4 Notes

1. Penicillin–streptomycin is optional for this cellular medium as with others. There is some discussion that the addition of antibiotics to the culture medium may fundamentally alter aspects of the cell membrane and can therefore impede or change results. This has not been objectively identified but may be assessed as a step for troubleshooting if experiments are unsuccessful. Furthermore, this is a standard culture medium for HEK293 and HeLa cell lines used in this protocol. Other media optimization may be necessary if other cell lines are used.

2. For culture of HL-1 murine atrial myocyte lines, this medium is essential for growth and maintaining contractility. Either 100 µM norepinephrine or 10 µM epinephrine may be used. The composition of this medium used in this protocol has been optimized with both epinephrine and norepinephrine in the past yielding contractile cells with appropriate cell–cell communication. Culture media may require further optimization if other atrial lines are used.

3. For automation, the microscope, microinjector, and micromanipulator can be linked together. The apparatus can use a centralized computer processor and software for tip detection, micromanipulation, and microinjection. This system is best highlighted in 2 papers by Liu and colleagues which first created this automation process. As the process of automation is complex, the specifics of the robotic setup including how to create and operate the system are beyond the scope of this chapter and can be found in those publications [20, 21].

4. Dye solution selection must be informed to the gap junction being assessed. A combination of dyes—a small molecular weight and a large molecular weight—is required in order to identify the injected cell and the surrounding cells to which it is connected. Knowledge of the gap junctions being assessed helps to inform which dyes to use based on the molecular weight restriction of the pore. Another commonly used dye in these experiments was lucifer yellow (MW = 457 Da).

5. Especially with contractile cell lines like HL-1, having young (i.e., low passage) cell lines is important. As cell lines age from passage to passage, the contractility of the line is reduced and a visual marker of gap junction connectivity is lost.

6. In a typical experiment, seeding to more than a 60 mm plate would be challenging unless using automation. 35 mm plates are sufficient because they have enough surface area in which to comfortably perform microinjection while also having enough cell mass to perform multiple experiments on a single plate.

7. For HL-1 cells, a healthy culture will have automaticity and contractility. Seeing patches of cells contracting together is a strong indicator of effective gap junction communication. For noncontractile cell lines, cell-to-cell contact should be visualized in order to assume communication. Do not allow cells to grow overconfluent as this will diminish the effectiveness of intercellular communication.

8. Avoid trypsinizing cells when performing passaging as this can damage the extracellular components of the gap junctions. Low concentration trypsin (i.e., 0.02 %) and physical agitation is sufficient to dislodge cells while minimally impacting the membrane surface proteins.

9. Dyes should be shielded from light exposure in order to minimize photobleaching and maximize signal. Wrapping tubes in foil can help.

10. This allows capillary forces to fill micropipette and minimizes bubbles in the tip.

11. Using a wire brush (usually used to wash graduated cylinders) with a braided metal handle and gently pulling it across the midsection of the micropipette can create sufficient vibration to dislodge bubbles.

12. If compensation pressures are too low, there will be insufficient injection into the cells. If too high, the background dye level will increase and the amount injected into cells will be inconsistent.

13. In previous experiments, a relationship between injection volume and injection pressure was found [21]. Most cytosolic

injections, to prevent cell death, must not exceed 5 % of cytosolic volume. In these experiments, an injection pressure of 3000 hPa would consistently provide about 30 fL of injection volume. This could be consistently and reproducibly used throughout the high-throughput robotic or manual dye injections.

14. Selecting an area of adequate cell growth and quality is important at this stage to ensure that injections occur in a region that can be found again on microscopy. If cells are selected that are too sparsely connected, the utility of the microinjection will be limited.

15. Clear views of the membranes are very important to the success of microinjection. It is important to see the dimpling of the membrane clearly to ensure that puncture occurs.

16. This requires patience. Adjust the microscope view over top of the cell of interest. Readjust the view until the micropipette tip is visible in the plane and use coarse or fine adjustments to place the micropipette tip above the cell of interest. Slowly lower the micropipette tip and adjust the microscope plane to view the descent of the micropipette and maintain position while also watching to ensure that the tip continues to expel small trickles of the dye mixture. Once the cell comes into view more clearly, adjust the microscope to the previous setting that best visualized the membrane while also allowing you to see the out-of-focus micropipette tip.

17. This will create a dimpling in the cell membrane. The dimpling will disappear when the micropipette punctures the membrane.

18. Prior to the official experiment, it is important to optimize the injection pressure, injection volume, and concentration of dye used. Modify the variables according to the cell line used and trial the methodology to optimize to the experiment's requirements. Not all settings will work well in all cells. This requires patience and careful consideration.

19. For shorter timeframe dye transfer assays, it is helpful to start a timer with your first injection and mark the time that each cell was injected. This way, consistency can be applied when waiting for each cell to perfuse. Furthermore, by drawing quadrants on the dish beforehand, specific time intervals can be decided upon before hand creating a setup for time-elapsed experimentation.

20. The number of repetitions is reliant on the experiment and statistical power required.

21. Unfortunately, puncturing the membranes may trigger apoptosis and limit the reliability of the data from those cells.

In most cases, the dye transfer assay can still be performed despite cell death but this should be taken into consideration. It is recommended to inject many cells for reliable results.

22. Ensure that injected cells are sufficiently distanced from each another so as to not have overlapping areas postinjection. If areas overlap, the quantification of dye transfer is limited.

23. Avoid spending too much time injecting dye into cells as the intensity of the light from the room and the microscope may decrease the amount of fluorescence the dye may produce.

24. For most dye transfer assays, 5–10 min is sufficient in these transfer assays. This time may vary based on experimental design. Too long times may result in dilution of the dye during spread. Short times may mean the full extent of spread is not identified.

Acknowledgements

Special thanks to the following individuals for their hard work in helping to develop this protocol and the experiments for the testing thereof: Vinayakumar Siragam, Zheng Gong, Jun Chen, Clement Leung, Zhe Lu, Changhai Ru, Shaorong Xie, Jun Luo, and Lynn Strandberg. This work was supported by the Canadian Institutes of Health Research Team grant for ARVC Research (2009–2014), CIHR/NSERC Collaborative Health Research Project (2015–2018), The Caitlin Elizabeth Morris Fund of Appliance Canada, The Alex Corrance Memorial Fund, and the University of Toronto Paediatric Research and Clinical Summer Scholarship.

References

1. Alexander DB, Goldberg GS (2003) Transfer of biologically important molecules between cells through gap junction channels. Curr Med Chem 10:2045–2058

2. Hertzberg EL, Lawrence TS, Gilula NB (1981) Gap junctional communication. Annu Rev Physiol 43:479–491

3. Heyman NS, Kurjiaka DT, Ek Vitorin JF et al (2009) Regulation of gap junctional charge selectivity in cells coexpressing connexin 40 and connexin 43. Am J Physiol Heart Circ Physiol 297:450–459

4. Pfenniger A, Wohlwend A, Kwak BR (2011) Mutations in connexin genes and disease. Eur J Clin Invest 41:103–116

5. Gutstein DE, Danik SB, Lewitton S et al (2005) Focal gap junction uncoupling and spontaneous ventricular ectopy. Am J Physiol Heart Circ Physiol 289:1091–1098

6. Strom M, Wan X, Poelzing S et al (2010) Gap junction heterogeneity as mechanism for electrophysiologically distinct properties across the ventricular wall. Am J Physiol Heart Circ Physiol 298:787–794

7. El-Fouly MH, Trosko JE, Chang CC (1987) Scrape-loading and dye transfer. Exp Cell Res 168:422–430

8. Raptis LH, Brownell HL, Firth KL et al (1994) A novel technique for the study of intercellular,

junctional communication: electroporation of adherent cells on a partly conductive slide. DNA Cell Biol 13:963–975

9. Wade MH, Trosko JE, Schindler M (1986) A fluorescence photobleaching assay of gap junction-mediated communication between human cells. Science 232:525–528

10. McCain ML, Desplantez T, Geisse NA et al (2012) Cell-to-cell coupling in engineered pairs of rat ventricular cardiomyocytes: relation between Cx43 immunofluorescence and intercellular electrical conductance. Am J Physiol Heart Circ Physiol 302:443–450

11. Fitzgerald DJ, Murray AW (1980) Inhibition of intercellular communication by tumor-promoting phorbol esters. Cancer Res 40:2935–2937

12. Zhang Y (2007) Microinjection technique and protocol to single cells. Protoc Exch. doi:10.1038/nprot.2007.487

13. Abbaci M, Barberi-Heyob M, Blondel W et al (2008) Advantages and limitations of commonly used methods to assay the molecular permeability of gap junctional intercellular communication. Biotechniques 45:33–62

14. Czyz J, Irmer U, Schulz G et al (2000) Gap-junctional coupling measured by flow cytometry. Exp Cell Res 255:40–46

15. Juul MH, Rivedal E, Stokke T et al (2000) Quantitative determination of gap junction intercellular communication using flow cytometric measurement of fluorescent dye transfer. Cell Adhes Commun 7: 501–512

16. Sun Y, Duthaler S, Nelson BJ (2004) Autofocusing in computer microscopy: selecting the optimal focus algorithm. Microsc Res Tech 65:139–149

17. Liu J, Gong Z, Tang K et al (2014) Locating end-effector tips in robotic micromanipulation. IEEE Trans Robot 30:125–130

18. Wang WH, Liu XY, Sun Y (2007) Contact detection in microrobotic manipulation. Int J Robot Res 26:821–828

19. Wang W, Sun Y, Zhang M et al (2008) A system for high-speed microinjection of adherent cells. Rev Sci Instrum 79:104302

20. Liu J, Siragam V, Gong Z et al (2014) Robotic adherent cell injection for characterizing cell-cell communication. IEEE Trans Biomed Eng 62:119–125

21. Liu J, Siragam V, Chen J et al (2014) High-throughput measurement of gap junctional intercellular communication. Am J Physiol Heart Circ Physiol 306:1708–1713

Chapter 11

Electroporation Loading and Dye Transfer: A Safe and Robust Method to Probe Gap Junctional Coupling

Elke Decrock, Marijke De Bock, Diego De Baere, Delphine Hoorelbeke, Nan Wang, and Luc Leybaert

Abstract

Intercellular communication occurring via gap junction channels is considered a key mechanism for synchronizing physiological functions of cells and for the maintenance of tissue homeostasis. Gap junction channels are protein channels that are situated between neighboring cells and that provide a direct, yet selective route for the passage of small hydrophilic biomolecules and ions. Here, an electroporation method is described to load a localized area within an adherent cell monolayer with a gap junction-permeable fluorescent reporter dye. The technique results in a rapid and efficient labeling of a small patch of cells within the cell culture, without affecting cellular viability. Dynamic and quantitative information on gap junctional communication can subsequently be extracted by tracing the intercellular movement of the dye via time-lapse microscopy.

Key words Intercellular communication, Connexin, Gap junction, Electroporation, Dye transfer, Time-lapse imaging

1 Introduction

In a multicellular organism, cells do not function on their own, but they highly interact with each other to provide a coordinated response to certain intracellular and extracellular conditions [1]. The interaction can exist of merely adhesive capacity, but it can also mediate the actual exchange of signaling molecules between neighboring cells via specialized contacts consisting of vast arrays of plasma membrane channels, called gap junction channels (GJCs) (Fig. 1) [2, 3]. These channels connect the cytoplasm of adjacent cells and arise from the head-to-head interaction of two hemichannels, being hexamers composed of connexin (Cx) subunits. The latter are tetraspan membrane proteins of which 21 human species have been identified and named according to their molecular weight (MW). They are present in most organs and display a tissue and cellular specificity with Cx43 being the most abundant and widespread Cx in mammals [4, 5].

Mathieu Vinken and Scott R. Johnstone (eds.), *Gap Junction Protocols*, Methods in Molecular Biology, vol. 1437,
DOI 10.1007/978-1-4939-3664-9_11, © Springer Science+Business Media New York 2016

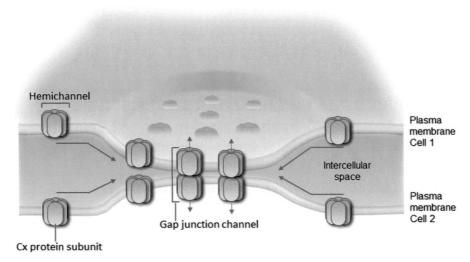

Fig. 1 *Schematic diagram of gap junctions and their subunits.* A set of 6 Cx proteins arrange themselves into a hemichannel configuration after which they are inserted in the cellular membrane. Two hemichannels that are located at a cell–cell contact region align to form a GJC. Multiple GJCs concentrate at so-called gap junction plaques, where they allow the bidirectional diffusion-driven transport of ions and small hydrophilic molecules (i.e., <1.5 kDa) between neighboring cells. This process, denoted by gap junctional intercellular communication, allows for a coordinated response of cells to certain stimuli and is a prerequisite for normal tissue homeostasis

Most cells of normal tissue, except for differentiated skeletal muscle cells, and freely circulating cells including erythrocytes and circulating lymphocytes, communicate through GJCs [6]. Gap junctional intercellular communication (GJIC) is driven by the passive diffusion of small (i.e., <1–1.5 kDa) hydrophilic molecules, such as glucose, glutamate, glutathione, cyclic adenosine monophosphate, adenosine 5′-triphosphate, inositol 1,4,5-trisphosphate, and ions, including Ca^{2+}, K^+, Na^+, and Cl^-. Although GJCs appear to be a rather general route for the exchange of these molecules as well as other substances such as fluorescent dyes, their biophysical permeation properties actually depend on the nature of the Cx species that form the channel [7–12]. There is furthermore a tremendous diversity in the assembly of GJCs, as they can be composed of either homogeneous or heterogeneous subunits resulting in variations in pore sizes [3]. Not only the size, but also the charge of the permeating molecule as well as the presence of high affinity binding sites for specific transjunctional metabolites within the channel pore are suggested to underlie these differences in permselectivity [13–15]. The Cx-subtype furthermore dictates the specific regulatory properties of these channels, such as by membrane voltage, the intracellular environment (i.e., cytoplasmic Ca^{2+} and pH), interaction with proteins, posttranslational modifications, and pharmacological agents [3, 16].

Several methods are currently available for investigating GJIC, all with their advantages and limitations [1, 17, 18]. The most simple, widely applicable, and also best-known approach consists of the introduction of a small dye into living cells and microscopically tracking its intercellular diffusion to neighboring cells. Several hydrophilic molecules covering a broad range of sizes (i.e., below 1 kDa) and charges can be used [8, 12]. In order to visualize subsequent dye spread in function of time, the reporter dye should fulfill several characteristics, namely: (1) it should be visible by transmitted light or fluorescence microscopy, (2) the fluorescence intensity should remain stable during recording, (3) the dye has to be GJC permeable, but also plasma membrane impermeable in order to avoid transfer through the extracellular space, (4) the choice of a specific GJC reporter dye should reflect the Cx under study as outlined earlier, (5) the dye should be able to freely diffuse throughout the cytoplasm, and (6) it cannot be toxic to the cells or change intracellular properties of the cells that can influence GJIC, such as modifying the intracellular Ca^{2+} concentration or pH. Frequently used fluorescent dyes for evaluating GJIC are Lucifer Yellow (MW 457 Da, charge -2), 6-carboxyfluorescein (6-CF; MW 376 Da, charge -2), calcein (MW 623 Da, -4), or Alexa Fluor dyes (available in the 350–760 Da MW range). Small nonfluorescent molecules, such as neurobiotin, are also frequently applied as tracers to probe GJIC. However, time-lapse imaging of dye transfer of these molecules is not possible, since the samples have to undergo additional fixation and staining procedures to visualize them. Of note, although channels composed of different Cx species display distinct permeability characteristics, most Cx channels are permeable to several tracers. Unfortunately, there are currently no dyes available that are specific for a channel composed of certain Cx subunits.

In principle, dye introduction can be achieved by several approaches, for example, by impaling cells with sharp pipettes for microinjection or by making a scrape in a cell monolayer using a blade or needle [19, 20]. However, these methods do not only damage cells but also modify or even washout certain intracellular components that are required for normal physiological signaling. Ideally, the loading technique should be minimally invasive and able to provide dynamic and quantitative information of dye transfer. Here, an electroporation technique is described that grants the loading of a spatially restricted area within an adherent cell population in vitro with a fluorescent GJC-permeable reporter dye. Subsequent time-lapse monitoring of dye diffusion throughout the cell population allows for the visualization of cells that are connected to each other through GJCs, a technique further denoted by electroporation loading and dye transfer (ELDT). Electroporation designates the application of high voltage pulses to create transient nanometer-scale pores into the plasma membrane via which extracellularly applied plasma membrane-impermeable molecules, such as nucleic acids, drugs, antibodies, proteins/peptides, radiotracers, and reporter dyes, can enter the cell [21–25]. Pore formation happens on a time scale of microseconds while closure occurs within seconds. By comparing electroporation with other dye loading techniques, such as microinjection and scrape loading, we conclude that it offers at least 4 advantages: (1) when optimized, it is a safe and superior method to evaluate GJIC as it is does not negatively affect the cellular physiology and viability [26–28]. The latter can have adverse effects on GJIC, e.g. via a perturbation of the cellular Ca^{2+} homeostasis. (2) It is a highly versatile technique that allows the labeling of small or large areas within a cell culture with virtually any compound with responses subsequently being monitored in adjacent nonloaded areas. We have successfully applied the technique for the loading of fluorescent dyes, caged compounds, molecules/peptides interfering with signaling pathways, and even recombinantly expressed and purified enzymes, and this in the context of Ca^{2+} signaling and cell death studies, and for measuring GJIC as outlined later [26, 28–38]. (3) The technique is not restricted to specific cell types [26, 28–39]. (4) Only minimal quantities of the electroporated compound are required for an efficient loading as outlined in the protocol later.

The loading method necessitates an electrical drive circuit providing an electroporation signal and an electrode to apply the signal to a localized zone in a cell monolayer [26, 40]. Ideally, the electroporation protocol should combine a good loading efficiency, while having only a minimal impact on cell viability in order to allow subsequent functional studies. Both the loading efficiency and the cell viability primarily depend on the applied electric pulse parameters [26, 41, 42]. The latter can be controlled in such a way that the pores have sufficient large radii to allow the entrance of extracellular molecules to the cells' interior and reclose rapidly without causing a disturbance in cellular physiology. We recently optimized an electroporation protocol to load a narrow strip of adherent cells with a GJC-permeable fluorescent reporter dye. We demonstrated that the application of a high-frequency (i.e., 50 kHz) electrical signal oscillating around the zero potential as a bipolar alternating current (AC) is preferred over a high amplitude direct current (DC) protocol to load the cells with the dye and to evaluate subsequent dye spread [26]. The main reason is that the absence of a DC component in the electrical drive signal preserves cell viability. The AC-coupled stimulation can be practically achieved by using an amplifier with a transformer to isolate the output of the signal-generating device [26, 27]. The 50 kHz voltage oscillations are applied in a pulsed and repeated manner with the oscillatory signal being switched-on ten times per s for 2 ms, forming a pulse train that is repeated 15 times with 0.5 s pauses (Fig. 2a).

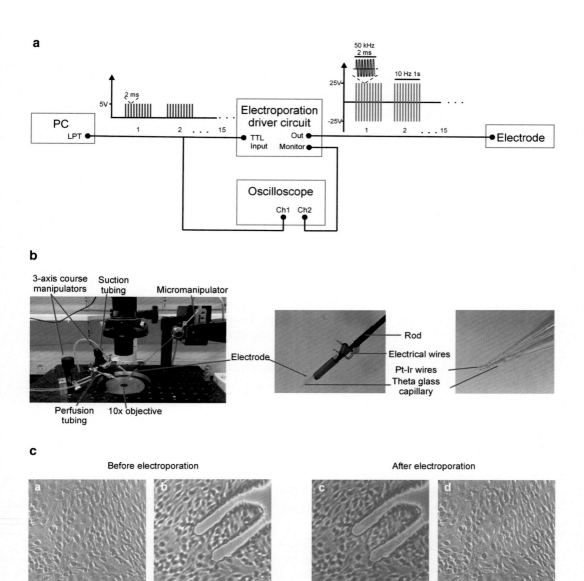

Fig. 2 *Overview of the electroporation setup used for the generation and registration of bipolar pulses.* (**a**) The computer generates a TTL pulse that serves as input for the electroporation driver circuit and as a trigger for the oscilloscope. The TTL pulse is converted by the electroporation driver circuit to a 50 kHz oscillating signal that alternates between positive and negative voltages, and that is sent to the electrode and is monitored by the oscilloscope. This oscillating signal is applied in 15 series of 1 s duration, each consisting of 10 repetitions of 2 ms duration. A schematic overview of the driver circuit has been published previously [26, 27] and a detailed diagram of the generator and amplifier setup is also available on request. (**b**) Picture of the microscope stage and the electrode. Note the two parallel Pt/Ir wires that are protruding out of the tip of the theta glass capillary. The exposed ends are about 300 µm in length (i.e., higher magnification in Fig. 2c). During the experimental procedure, the electrode is positioned above the cells using the micromanipulator. (**c**) Phase contrast/fluorescence overlay images taken before (i.e., without (*a*) and with (*b*) the electrode positioned on top of the cells) and directly after electroporation loading (i.e., after removal of the remaining extracellular dye solution) (*c–d*). Cell morphology and viability are well preserved after electroporation. Note that in (*b*) and (*c*) the electrode is positioned 100 µm above the cell layer. The objective is positioned in the focal plane of the electrode and 100 µm higher relative to the focal plane of the cell layer. The scale bar measures 100 µm

Loading of a restricted area can be achieved by applying the bipolar voltage pulses to a 2-wire electrode that is positioned in close proximity to the target cells. The area of electroporated cells depends on the electrode dimensions. In previous studies, we used an electrode consisting of 2 parallel Pt/Ir wires with a diameter of 120 μm, 8 mm long and separated by 500 μm. When placed about 400 μm above the cells and applying a peak-to-peak oscillation voltage of 100 V, this results in a longitudinal electroporated zone of about 300 μm wide and 8 mm long (i.e., surface area about 2.4 mm^2) [26–28, 34]. The actual size of the electrode can be modified according to specific needs. In a first section of this chapter, we describe the assembly of a 2-wire fork-shaped electrode instead of the previously strip-shaped, using 50 μm diameter Pt/Ir wire which results in the electroporation of a smaller patch of cells with an average surface area of about 0.07 mm^2 (Fig. 2b, c). In a second section, a detailed procedure for the time-lapse microscopy imaging as well as the analysis of dye spread to neighboring nonloaded cells is provided.

2 Materials

2.1 Assembly of a Parallel 2-Wire Fork-Shaped Pt/Ir Electrode

1. Borosilicate theta capillary glass (OD/ID 2/1.4 mm, length 10 cm; Harvard apparatus, USA).

2. Pt/Ir wire 90:10, diameter 50 μm (Advent Research Materials Ltd, England).

3. Microelectrode puller (Model P-97, Sutter Instrument, USA).

4. Micromanipulator (PatchMan NP2, Eppendorf, Belgium).

2.2 Electroporation Loading and Time-Lapse Imaging of Dye Transfer

1. Adherent confluent monolayer culture (see **Note 1**) seeded on polystyrene dishes or glass bottom dishes (CELLview™ Dish, Greiner Bio One, Belgium).

2. Electroporation buffer: 300 mM sorbitol, 4.02 mM KH$_2$PO$_4$, 10.8 mM K$_2$HPO$_4$, 1.0 mM MgCl$_2$, and 2.0 mM 4-(2-hydroxyethyl)-1-piperazineethanesulfonic acid (HEPES). Adjust to pH 7 and store at 4–8 °C.

3. Electroporation solution: 1 mM 6-CF (Life Technologies, Belgium) and 1 mM 10 kDa Dextran Rhodamine (DR) (Life Technologies, Belgium) in electroporation buffer (see **Notes 2–4**). Aliquots can be stored at –20 °C and should be thawed in the absence of light. Repetitive thawing should be avoided. The solution should be at room temperature *prior* to use.

4. Hanks Balanced Salt Solution (HBSS-HEPES): 0.81 mM MgSO$_4$·7H$_2$O, 0.95 mM CaCl$_2$·2H$_2$O, 137 mM NaCl, 0.18 mM Na$_2$HPO$_4$·2H$_2$O, 5.36 mM KCl, 0.44 mM KH$_2$PO$_4$, 5.55 mM d-glucose, and 25 mM HEPES. Adjust to pH 7.4 and store at 4–8 °C.

5. 10 μL microloader pipette tips.

6. Electroporation driver circuit (*see* **Note 5**).

7. Oscilloscope for inspection of peak-to-peak electroporation voltage (ISO-TECH IDS 6072A-U, RS Components, Belgium).

8. Computer to generate Transistor–Transistor Logic (TTL) pulses.

9. Nikon Eclipse TE300 inverted epifluorescence microscope (Nikon Belux, Belgium) positioned on an antivibration table and supplied with an EM-CCD camera (QuantEM™ 512SC CCD camera, Photometrics, USA), a Lambda DG-4 filterswitch (Sutter Instrument Company, Novato, Canada), a multiband dichroic mirror and emitter filter set (XF2050 and XF3063, respectively, Omega Optical, USA), and a 10× objective (Plan APO, NA 0.45—Nikon). The microscopic stage is equally equipped with the electroporation electrode mounted on a micromanipulator and suction and perfusion tubes, positioned in 3-axis course manipulators and connected, respectively, to a suction device and perfusion system (Fig. 2b) (*see* **Note 6**).

10. Quantem Frames imaging software (custom developed in Microsoft Visual C⁺⁺ 6.0).

2.3 Analysis of Dye Spread

FluoFrames analysis software (custom developed in Microsoft Visual C⁺⁺ 6.0) (*see* **Note 7**).

3 Methods

3.1 Assembly of a Parallel 2-Wire Fork-Shaped Electrode

Parallel bipolar microelectrodes are commercially available, e.g. at FHC (Canada). Alternatively, the electrode can be fabricated according to the procedure described as follows.

1. Pull glass pipettes and break the tip to obtain a tip size of approximately 150 μm outer diameter (Fig. 2b).

2. Insert a 5 cm long Pt/Ir wire into each lumen of the theta glass pipette (*see* **Note 8**). Using a blade, shorten the protruding wires at the tip of the pipette under microscopy observation such that the exposed ends are about 300 μm in length (Fig. 2c).

3. Fix the wires into the glass pipette by applying a small amount of cyanoacrylate glue to the tip of the capillary to seal around each wire.

4. Solder each Pt/Ir wire at the other end to an electric wire for connection to the driver box.

5. Attach the pipette to a rod for subsequent mounting on a micromanipulator and connection to the driver circuit.

3.2 Electroporation Loading and Time-Lapse Imaging of Dye Transfer

1. Preheat the HBSS-HEPES and the low-conductivity electroporation buffer to 37 °C to avoid a sudden temperature drop upon medium change.

2. Prepare the electroporation buffer solution containing the desired compounds.

3. Remove the cell culture dish from the incubator and place it on the stage of the microscope.

4. Rinse the cells three times with HBSS-HEPES, followed by two times with the low-conductivity electroporation buffer solution (*see* **Note 9**) using the suction/perfusion system.

5. Completely remove the electroporation buffer solution (*see* **Note 10**).

6. Visualize and focus to the cells using transmitted light microscopy and subsequently move the objective 100 µm higher relative to the focal plane of the cell layer via the calibrated micrometer dial on the focus knob.

7. Under transmitted light microscopy observation and using the micromanipulator, guide the electrode toward the area of interest until the wires are in perfect focus at the edges.

8. Administer 10 µL of the electroporation solution at the tip of the electrode with a microloader tip and apply the AC-coupled electroporation drive signal to the electrode (*see* **Note 11**).

9. Rinse the cells with HBSS-HEPES using the suction/perfusion system.

10. Start imaging. Images are recorded at room temperature and in the absence of light at a 5 min time interval, and this for a 25 min period post-electroporation. The cells are kept in HBSS-HEPES during the whole imaging period. Simultaneous imaging of 6-CF and 10 kDa DR is carried out by exciting the dyes using 482 and 568 nm light, respectively, that is obtained from a Xenon light source and a Lambda DG-4 filterswitch. Emission passes through a multiband dichroic mirror and emitter filter set.

3.3 Quantification of Dye Spread

Intercellular dye spread can be quantified by counting the number of cells that received the GJC-permeable dye from the initially loaded cells. An alternative approach consists of calculating the surface area of dye spread and the half-maximal dye transfer index as outlined as follows.

1. Apply a threshold to all images. The threshold is derived from the histogram of the pixel intensity distribution of the image taken immediately after electroporation (0′ time point in Figs. 3 and 4) and corresponds to the upper level of the background noise. The latter typically generates a large peak at lower intensity values. We set the threshold at the right flank of the peak at half-maximal pixel intensity.

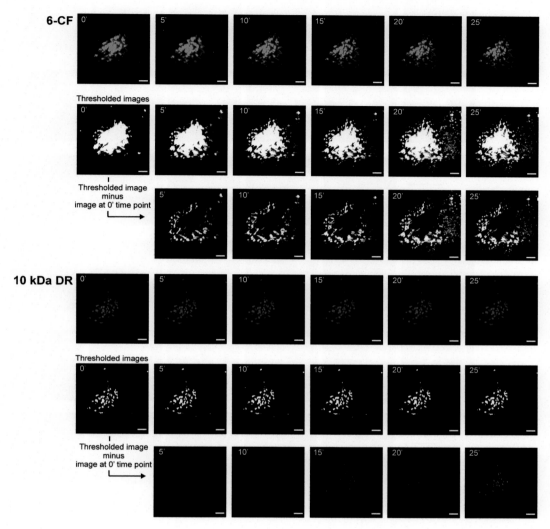

Fig. 3 *Dye transfer experiment in C6 glioma cells stably transfected with Cx43.* C6 glioma cells were electroporated with a solution containing the GJC-permeable 6-CF and the high molecular weight dye 10 kDa DR. Images were taken at several time points after electroporation, were thresholded, and were subsequently subtracted by the image taken immediately after electroporation (i.e., 0′ time point). The resulting image displays the dye transfer area that contains cells (in *white*) that received dye from the electroporated area. Quantitative information is presented in Fig. 4. The scale bar measures 100 μm

2. Subtract the first image, recorded immediately after electroporation (0′ time point in Figs. 3 and 4) on a pixel-to-pixel basis from all subsequent images.

3. Calculate the area of above threshold pixels and convert to a μm² scale. The surface area can be represented as a function of time from which the half-maximal dye transfer rate index can be determined (Figs. 3 and 4).

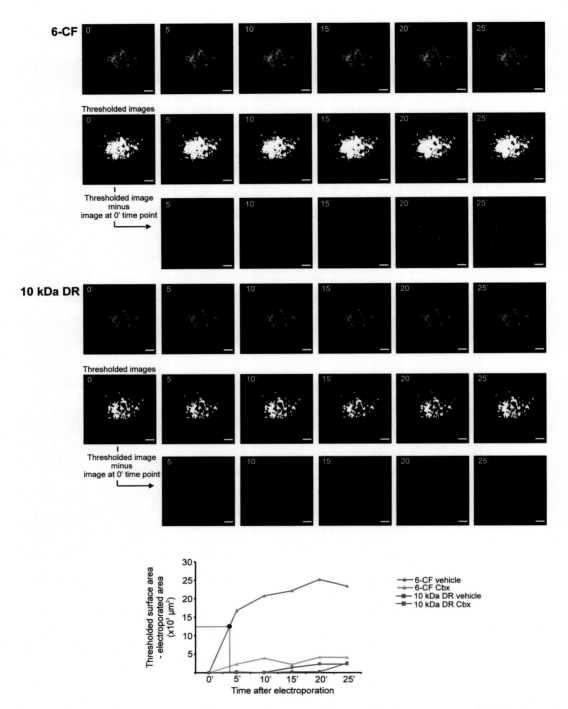

Fig. 4 *Dye transfer experiment in C6 glioma cells stably transfected with Cx43 in the presence of the Cx channel blocker carbenoxolone (Cbx).* Similar experiment as in Fig. 3, but now 25 μM Cbx was added to the supernatant 15 min before electroporation and also during the 25 min imaging period. The graph represents the dye transfer area in function of time after electroporation. In the absence of Cbx, the surface area of the 6-CF dye expanded over time, whereas the 10 kDa DR did not spread to neighboring cells, but remained locally in the electroporated area. The presence of Cbx strongly decreased the spread of 6-CF to neighboring cells. The half-maximal dye transfer rate index (i.e., the ratio of the half-maximal surface area to the corresponding time) for 6-CF can be derived from the graph (*black dot*) and is equal to 59.5 μm^2 per s. The scale bar measures 100 μm

4   Notes

1. We have used the electroporation method to load various substances in different cell types including C6 glioma cells, endothelial cell lines (RBE4 and bEND.3), epithelial cell lines (MDCK, ECV304 and HeLa cells), fibroblast cell lines (MEF and HEK), and primary cortical astrocytes [26, 28–39].

2. Figures 3 and 4 demonstrate the transfer of the GJC-permeable dye 6-CF through GJCs composed of Cx43 in a C6 glioma culture stably transfected with Cx43 in the absence or presence of the Cx channel blocker carbenoxolone (Cbx). However, as outlined, the permeability of GJCs for specific dyes is related to the Cx subunit composition. The choice of the GJC reporter dye thus depends on the Cx under study.

3. A second high MW dye can be co-electroporated to test for dye transfer not mediated by GJCs. Obviously, the excitation/emission of this reporter dye may not coincide with the excitation/emission spectrum of the GJIC indicator. Commonly applied GJ-impermeable fluorescent reporter markers include 10 kDa DR or 10 kDa FITC-dextran.

4. Caution should be taken when loading agents that are dissolved in DMSO concentrations above 0.2 %. Although it has been described that DMSO (i.e., 1.25 %) stabilizes the plasma membrane during electroporation loading and improves loading efficiency [43, 44], it might affect GJIC [45]. It is also recommended to filter the electroporation solution in order to remove any particulates.

5. The device to generate a 50 kHz AC-coupled pulsed electroporation signal is nonstandard equipment and is not commercially available. A schematic overview of the driver circuit has been published previously [26, 27] and a detailed diagram of the generator and amplifier setup is also available on request.

6. In order to perform time-lapse imaging of the gap junctional dye spread, it is imperative to have the microscope used for positioning the electrode equipped with a light source and filters to excite the fluorescent dyes as well as a suction/perfusion device to remove residual dye after electroporation. Since the electroporation procedure has to take place in the absence of any vibrations, it is convenient to have the microscope positioned on an antivibration table. Finally, dye transfer to neighboring cells can be observed, provided that the level of illumination is kept low. In order to avoid cellular damage from illumination, it is advised to use a sensitive high-quality CCD camera. Here, imaging is carried out using an electron multiplying QuantEM™ 512SC CCD camera from Photometrics (USA).

7. The analysis can also be performed with Adobe Photoshop software (Adobe Systems, USA) or Image J software (http://imagej.nih.gov/ij/).

8. A theta glass pipette is used to protect and separate the thin wires of the bipolar electrode.

9. This low conductivity buffer is specifically formulated to minimize current flow during the electroporation procedure, thus preventing any significant damage to the cells.

10. From this stage on, the cells only contain some solution that is kept there by capillary forces (i.e., clefts between cells) and attraction by osmotic forces. The next steps (i.e., 6–9) need to be performed as rapidly as possible.

11. The electroporation protocol consists of 15 pulse trains, each containing 10 pulses (i.e., 2 ms in duration) per second of a 50 kHz electrical signal and separated by 0.5 s breaks (Fig. 2a). Important to note is that (1) the concentration that a compound reaches within the cell after electroporation loading is lower to that within the electroporation solution and (2) the loading efficiency decreases with higher MW. A loading efficiency of about 40 % can be obtained with a 50 V peak-to-peak voltage applied to electrode wires separated by 150 μm (i.e., electrical field strength ~3300 V/cm) for molecules with a MW of 0.4–10 kDa [26]. In addition, with these settings, cell death in the electroporated area is not different from spontaneous cell death in nonelectroporated zones of the cell cultures as measured by uptake of the cell impermeable and DNA integrating dye propidium iodide (i.e., 30 μm, 15 min incubation at room temperature) at 5 min postelectroporation (i.e., 0.56 % ±0.15 in the electroporated area versus 0.47 % ±0.21, respectively, in surrounding nonelectroporated cells ($n=3$)) [26, 28]. However, a careful and thorough optimization of the electroporation parameters is required for each cell type used as the efficiency and safety (i.e., cell death) of a given field strength and electrode morphology critically depends on the geometry of the cells [46]. Optimization is done by varying the distance of the electrode from the cells and the voltage applied to the electrode until optimal loading efficiency and minimal cell death is achieved. An electrode that is positioned too close to the cells or a too high voltage will result in increased cell death counts. In contrast, an electrode that is positioned too far from the cells or a too low voltage will provide a low loading efficiency and thus less dye in the cell's interior to be spread to neighboring cells via GJCs. The loading efficiency can be estimated by calculating the ratio of the fluorescence intensity of the reporter dye in the electroporation zone to the fluorescence intensity of a thin layer of the same marker, sandwiched between two glass coverslips. The thickness of this

layer should approximate the thickness of the monolayer of the cells used which can be estimated by recording Z-stacks of confocal laser scanning images of cells loaded with a fluorescent indicator dye. We refer the reader to our previous papers [27, 40] for a detailed protocol. Finally, in order to maintain a consistent loading efficiency, it is important to clean the electrode on a frequent basis with deionized water and from time to time with 1 mM NaOH.

Acknowledgements

This work is supported by the Fund for Scientific Research Flanders (FWO-Vlaanderen), Belgium (grants G.0298.11, G.0571.12, G.0A54.13, and G.0320.15), the Special Research Fund (BOF) of Ghent University (grant 01IO8314), and the Interuniversity Attraction Poles Program (Belgian Science Policy, project P7/10).

References

1. Wei Q, Huang H (2013) Insights into the role of cell-cell junctions in physiology and disease. Int Rev Cell Mol Biol 306:187–221

2. Herve JC, Derangeon M (2013) Gap-junction-mediated cell-to-cell communication. Cell Tissue Res 352:21–31

3. Maeda S, Tsukihara T (2011) Structure of the gap junction channel and its implications for its biological functions. Cell Mol Life Sci 68:1115–1129

4. Sohl G, Willecke K (2003) An update on connexin genes and their nomenclature in mouse and man. Cell Commun Adhes 10:173–180

5. Laird DW (2006) Life cycle of connexins in health and disease. Biochem J 394:527–543

6. Reinecke H, Minami E, Virag JI et al (2004) Gene transfer of connexin43 into skeletal muscle. Hum Gene Ther 15:627–636

7. Alexander DB, Goldberg GS (2003) Transfer of biologically important molecules between cells through gap junction channels. Curr Med Chem 10:2045–2058

8. Elfgang C, Eckert R, Lichtenberg-Frate H et al (1995) Specific permeability and selective formation of gap junction channels in connexin-transfected HeLa cells. J Cell Biol 129:805–817

9. Bukauskas FF, Elfgang C, Willecke K et al (1995) Heterotypic gap junction channels (connexin26-connexin32) violate the paradigm of unitary conductance. Pflugers Arch 429:870–872

10. Kanaporis G, Mese G, Valiuniene L et al (2008) Gap junction channels exhibit connexin-specific permeability to cyclic nucleotides. J Gen Physiol 131:293–305

11. Goldberg GS, Moreno AP, Lampe PD (2002) Gap junctions between cells expressing connexin 43 or 32 show inverse permselectivity to adenosine and ATP. J Biol Chem 277:36725–36730

12. Kanaporis G, Brink PR, Valiunas V (2010) Gap junction permeability: selectivity for anionic and cationic probes. Am J Physiol 300:C600–C609

13. Veenstra RD (1996) Size and selectivity of gap junction channels formed from different connexins. J Bioenerg Biomembr 28:327–337

14. Bevans CG, Kordel M, Rhee SK et al (1998) Isoform composition of connexin channels determines selectivity among second messengers and uncharged molecules. J Biol Chem 273:2808–2816

15. Bevans CG, Harris AL (1999) Direct high affinity modulation of connexin channel activity by cyclic nucleotides. J Biol Chem 274:3720–3725

16. Verselis VK, Srinivas M (2013) Connexin channel modulators and their mechanisms of action. Neuropharmacology 75:517–524

17. Abbaci M, Barberi-Heyob M, Blondel W et al (2008) Advantages and limitations of commonly used methods to assay the molecular permeability of gap junctional intercellular communication. Biotechniques 45:33–52

18. Maes M, Crespo Yanguas S, Willebrords J et al (2016) Models and methods for in vitro testing of hepatic gap junctional communication. Toxicol In Vitro 25:569–577

19. El-Fouly MH, Trosko JE, Chang CC (1987) Scrape-loading and dye transfer. A rapid and simple technique to study gap junctional intercellular communication. Exp Cell Res 168:422–430

20. Liu J, Siragam V, Chen J et al (2014) High-throughput measurement of gap junctional intercellular communication. Am J Physiol Heart Circ Physiol 306:H1708–H1713

21. Gehl J (2003) Electroporation: theory and methods, perspectives for drug delivery, gene therapy and research. Acta Physiol Scand 177:437–447

22. Teruel MN, Meyer T (1997) Electroporation-induced formation of individual calcium entry sites in the cell body and processes of adherent cells. Biophys J 73:1785–1796

23. Ho SY, Mittal GS (1996) Electroporation of cell membranes: a review. Crit Rev Biotechnol 16:349–362

24. Weaver JC (1993) Electroporation: a general phenomenon for manipulating cells and tissues. J Cell Biochem 51:426–435

25. Kotnik T, Frey W, Sack M et al (2015) Electroporation-based applications in biotechnology. Trends Biotechnol 33:480–488

26. De Vuyst E, De Bock M, Decrock E et al (2008) In situ bipolar electroporation for localized cell loading with reporter dyes and investigating gap junctional coupling. Biophys J 94:469–479

27. Decrock E, De Bock M, Wang N et al (2015) Electroporation loading and flash photolysis to investigate intra- and intercellular Ca²⁺ signaling. Cold Spring Harb Protoc 2015:239–249

28. Decrock E, De Vuyst E, Vinken M et al (2009) Connexin 43 hemichannels contribute to the propagation of apoptotic cell death in a rat C6 glioma cell model. Cell Death Differ 16:151–163

29. Braet K, Aspeslagh S, Vandamme W et al (2003) Pharmacological sensitivity of ATP release triggered by photoliberation of inositol-1,4,5-trisphosphate and zero extracellular calcium in brain endothelial cells. J Cell Physiol 197:205–213

30. Braet K, Mabilde C, Cabooter L et al (2004) Electroporation loading and photoactivation of caged InsP3: tools to investigate the relation between cellular ATP release in response to intracellular InsP3 elevation. J Neurosci Methods 132:81–89

31. Braet K, Paemeleire K, D'Herde K et al (2001) Astrocyte-endothelial cell calcium signals conveyed by two signalling pathways. Eur J Neurosci 13:79–91

32. Monaco G, Decrock E, Akl H et al (2012) Selective regulation of IP₃-receptor-mediated Ca²⁺ signaling and apoptosis by the BH4 domain of Bcl-2 versus Bcl-Xl. Cell Death Differ 19:295–309

33. De Bock M, Wang N, Bol M et al (2012) Connexin-43 hemichannels contribute to cytoplasmic Ca²⁺ oscillations by providing a bimodal Ca²⁺-dependent Ca²⁺-entry pathway. J Biol Chem 287:12250–12266

34. Decrock E, Krysko DV, Vinken M et al (2012) Transfer of IP₃ through gap junctions is critical, but not sufficient, for the spread of apoptosis. Cell Death Differ 19:947–957

35. Vervliet T, Lemmens I, Vandermarliere E et al (2015) Ryanodine receptors are targeted by anti-apoptotic Bcl-XL involving its BH4 domain and Lys87 from its BH3 domain. Sci Rep 5:9641

36. Monaco G, Decrock E, Arbel N et al (2015) The BH4 domain of anti-apoptotic Bcl-XL, but not that of the related Bcl-2, limits the voltage-dependent anion channel 1 (VDAC1)-mediated transfer of pro-apoptotic Ca²⁺ signals to mitochondria. J Biol Chem 290:9150–9161

37. Vervliet T, Decrock E, Molgo J et al (2014) Bcl-2 binds to and inhibits ryanodine receptors. J Cell Sci 127:2782–2792

38. Monaco G, Decrock E, Nuyts K et al (2013) Alpha-helical destabilization of the Bcl-2-BH4-domain peptide abolishes its ability to inhibit the IP₃ receptor. PLoS One 8:e73386

39. Leybaert L, Sanderson MJ (2001) Intercellular calcium signaling and flash photolysis of caged compounds: a sensitive method to evaluate gap junctional coupling. Methods Mol Biol 154:407–430

40. Decrock E, De Bock M, Wang N et al (2015) Electroporation loading of membrane-impermeable molecules to investigate intra- and intercellular Ca²⁺ signaling. Cold Spring Harb Protoc 2015:284–288

41. Wegener J, Keese CR, Giaever I (2002) Recovery of adherent cells after in situ electroporation monitored electrically. Biotechniques 33:348, 350, 352 passim

42. Kotnik T, Pucihar G, Rebersek M et al (2003) Role of pulse shape in cell membrane electropermeabilization. Biochim Biophys Acta 1614:193–200

43. Fyrberg A, Lotfi K (2010) Optimization and evaluation of electroporation delivery of siRNA

in the human leukemic CEM cell line. Cytotechnology 62:497–507

44. Melkonyan H, Sorg C, Klempt M (1996) Electroporation efficiency in mammalian cells is increased by dimethyl sulfoxide (DMSO). Nucleic Acids Res 24:4356–4357

45. Yoshizawa T, Watanabe S, Hirose M et al (1997) Dimethylsulfoxide maintains intercellular communication by preserving the gap junctional protein connexin32 in primary cultured hepatocyte doublets from rats. J Gastroenterol Hepatol 12:325–330

46. Cegovnik U, Novakovic S (2004) Setting optimal parameters for *in vitro* electrotransfection of B16F1, SA1, LPB, SCK, L929 and CHO cells using predefined exponentially decaying electric pulses. Bioelectrochemistry 62:73–82

Chapter 12

Using Fluorescence Recovery After Photobleaching to Study Gap Junctional Communication In Vitro

Maria Kuzma-Kuzniarska, Clarence Yapp, and Philippa A. Hulley

Abstract

Fluorescence recovery after photobleaching (FRAP) is a microscopy-based technique to study the movement of fluorescent molecules inside a cell. Although initially developed to investigate intracellular mobility, FRAP can be also used to measure intercellular dynamics. This chapter describes how to perform FRAP experiment to study gap junctional communication in living cells. The procedures described here can be carried out with a laser-scanning confocal microscope and any in vitro cultured cells known to communicate via gap junctions. In addition, the method can be easily adjusted to measure gap junction function in 3D cell cultures as well as ex vivo tissue.

Key words Fluorescence recovery after photobleaching, Confocal microscopy, Gap junctions, Intercellular communication, Calcein, In vitro, Monolayer, Tenocytes, Fluorescence recovery curve, Mobile fraction percentage

1 Introduction

Fluorescence recovery after photobleaching (FRAP) permits quantitative assessment of the mobility of fluorescent molecules. FRAP is used to address questions related to protein localisation and interactions, mobility of molecules within cell compartments as well as organelle dynamics. The principle behind FRAP is simple, fluorescently labeled molecules are irreversibly bleached in a well-defined area, called the region of interest (ROI). Following the bleaching, unbleached molecules from outside the ROI, diffuse into the bleached area. This movement is observed as an increase in fluorescence intensity (fluorescence recovery) in the ROI. The rate of fluorescence recovery reflects the mobility of molecules and the analysis of the recovery curve yields quantitative information about the process [1, 2].

Although FRAP was initially developed to study intracellular dynamics, Wade et al. adjusted the technique to measure the transfer of fluorescent molecules between cells. They extended the

Mathieu Vinken and Scott R. Johnstone (eds.), *Gap Junction Protocols*, Methods in Molecular Biology, vol. 1437,
DOI 10.1007/978-1-4939-3664-9_12, © Springer Science+Business Media New York 2016

analysis from a small region to the whole cell and showed that it is possible to perform multiple fluorescence measurements of the same living cell. By choosing a fluorescent dye that can easily be transferred through gap junctions, they were able to assess the movement of fluorescent molecules between cells and as a result created a new technique to study gap junctional communication [3]. One of the main advantages of FRAP over other methods used to study gap junction function, such as scrape loading and micro-injection, is that FRAP experiments are noninvasive, thus all procedures are performed without disrupting cell integrity. FRAP is also regarded as less time consuming when compared to microinjection, and less technically demanding when compared to dual whole-cell patch clamp [4].

A FRAP experiment to study gap junctional communication can be divided into three stages (Fig. 1). First, prebleach imaging is performed in order to gain information about the initial fluorescence level. Next the cell of interest is selected and bleached with a high-powered laser. Finally, post-bleach fluorescence recovery is monitored using the same laser but at low power. It is important to

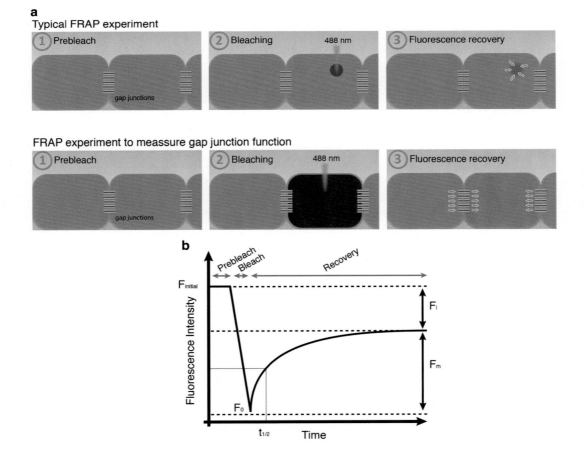

note that for the recovery to occur, the cells not only need to display functional gap junctions but enough cell partners must be present to exchange the fluorescent molecules (*see* **Note 1**) [5]. The fluorescent labeling can be performed using any fluorescent dye that can be transferred via gap junctions. Among different tracers calcein acetoxymethyl (AM) ester can be recommended. Calcein-AM is converted to calcein after acetoxymethyl ester hydrolysis by intracellular esterases. Upon the conversion to membrane impermeable calcein emits green fluorescence. Calcein is well retained by viable cells, yields high fluorescence, and can be effectively transferred between cells through gap junctions [6].

This chapter outlines how to set up, perform, and analyze FRAP experiments in cultured cells to assess gap junction function.

2 Materials

2.1 Cell Culture

1. Human hamstring tendon cells (tenocytes) isolated by standard explant culture method (*see* **Note 2**).

2. Cell culture media: Dulbecco's Modified Eagle Medium supplemented with 10% fetal bovine serum (FBS). Store at 2–8 °C.

3. Glass bottom dishes 35-mm with #1.5H (170 μm ± 5 μm) D 263 M Schott glass.

4. Incubator at 37 °C, 5% CO_2, 90% humidity.

5. Water bath at 37 °C.

2.2 Labeling with Calcein-AM

1. 1 mM calcein acetoxy-methylester (AM) stock solution dissolved in DMSO.

2. Serum-free culture medium: Dulbecco's Modified Eagle Medium with no supplements.

Fig. 1 A schematic representation of the principle of FRAP experiments. (**a**) In a typical FRAP experiment a population of intracellular molecules is fluorescently labeled. Next the fluorescent molecules in a small area, called the region of interest (ROI), are permanently bleached using a high power laser beam. Finally fluorescence recovery to the ROI is measured over time. The recovery in the ROI is based on the diffusion of the unbleached molecules into the bleached region of interest and provides information about the mobility of fluorescence molecules. In experiments employing the FRAP method to study gap junctions all fluorescently labeled molecules within a cell are permanently bleached, thus the region of interest is extended to the whole cell and the exchange of molecules via gap junctions between a bleached cell and the unbleached neighboring cells is responsible for the fluorescence recovery. (**b**) Two parameters can be used to describe fluorescence recovery in a bleached cell: recovery time and mobile fraction. $F_{initial}$—fluorescence intensity before bleaching, F_0—fluorescence intensity immediately after bleaching, F_m—mobile fraction (fraction that contributes to the recovery), F_i—immobile fraction (fraction that does not contribute to the recovery) $t_{1/2}$—half time recovery

3 Methods

3.1 Cell Culture

1. Grow tenocytes in cell culture medium.

2. Passage cells by scraping at 70 % confluency.

3. Seed tenocyte cells at a concentration of approximately 2×10^5 cells per well in 2 mL of medium onto 35 mm glass bottom dishes and culture to provide confluent cultures within 2–3 days (*see* **Note 3**).

3.2 Labeling with Calcein-AM

1. Thaw calcein-AM stock solution and prepare a 4 μM working solution by diluting 1:250, e.g., 4 μL of calcein-AM stock solution per 1 mL of warmed serum-free culture medium (*see* **Note 4**).

2. Remove media from tenocytes and wash cells once with warmed serum-free culture medium.

3. Remove the serum-free culture medium and replace with 1 mL of calcein-AM working solution.

4. Incubate the cells with calcein-AM for 15 min at room temperature or alternatively place the dishes in the incubator at 37 °C.

5. Cover to protect from light.

6. Wash the cells three times with pre-warmed serum-free medium to ensure that no calcein-AM is present in the medium.

7. Fill the dish with 1 mL of pre-warmed serum-free medium and proceed immediately with FRAP (*see* **Note 5**).

3.3 Setup for Fluorescence Recovery After Photobleaching

1. Perform procedure on a confocal laser scanning microscope (e.g., Zeiss LSM7 with Zen 2009 software) equipped with a 20×0.5 numerical aperture objective, a 488 nm argon-ion laser and an incubator chamber capable of maintaining temperature at 37 °C.

2. Mount the glass bottom dish with labeled cells on a motorized scan stage with an incubator chamber maintaining 37 °C.

3. Focus the microscope on the sample using epi-fluorescence mode with a GFP or similar filter cube.

4. Acquire one pre-bleach scan at low power to record the initial fluorescence intensity of the cell.

5. In order to accomplish complete photobleaching of calcein, adjust settings to irradiate the ROI four times using a pixel dwell time of 50 μs with argon-ion laser power set to the maximum (*see* **Note 6**).

6. Immediately after photobleaching the ROI, acquire a rapid time series (e.g., 5 s intervals for 4 min) to record the recovery of calcein. Ensure that the duration and laser power is set so that complete recovery is observed without excess photobleaching (*see* **Notes 7–9**).

7. Once appropriate conditions for photobleaching and imaging have been established, use the saved settings to perform FRAP experiments.

3.4 Performing the FRAP Measurement

1. Focus the microscope on the sample using epi-fluorescence mode with a GFP or similar filter cube.

2. Choose an appropriate region for photobleaching by manually drawing a ROI around selected cells.

3. Draw additional reference regions of interest around a cell not selected for bleaching and a region outside the cells to provide measures for non-bleached and background signal.

4. Photobleach the ROI using the previously established parameters.

5. Following photobleaching, record a rapid time series (e.g., 5 s intervals for 4 min) to record the recovery of calcein of fluorescence recovery using 1 % of the laser power (*see* **Note 10**). An example of a FRAP experiment and a representative fluorescence recovery curve can be found in Fig. 2.

6. Using software settings, the detector gain can be adjusted to image cells with varying labeling intensities.

7. It is possible to inhibit the recovery by using gap junction inhibitors (*see* **Note 11**) [5, 7].

8. Common problems associated with FRAP experiments, based on the type of recovery curve obtained in the experiment, are shown in Fig. 3.

9. Collect data for at least ten cells per treatment for each experiment.

10. In order to obtain reproducible data it is important to keep the basic imaging settings the same in all subsequent experiments and to not adjust bleaching and recovery conditions between experiments.

3.5 FRAP Analysis

1. Analysis of fluorescence recovery in bleached cells can be assessed by calculating the mobile fraction (F_m) percentage.

2. Software analysis, e.g., (Zeiss, ZEN 2009) can be used to identify the fluorescence intensities of the cell at full recovery (F_m), immediately after photobleaching (F_0), and before photobleaching ($F_{initial}$).

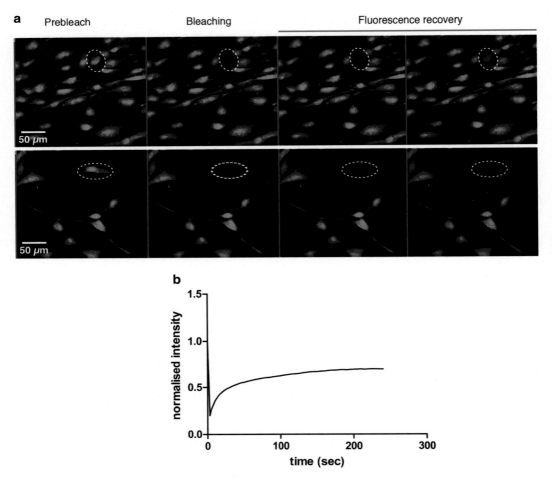

Fig. 2 An example of a FRAP experiment using human tendon cells (tenocytes). (**a**) Calcein labeled tenocytes were subjected to photobleaching using a 488 nm laser at 100 % power. Subsequently, a time-lapse series was recorded using 1 % power. The recovery is shown in images at <1 min and >1 min post-bleaching. As illustrated above, recovery depends on the number of adjacent cells and also the number of connections formed between cells. Cells grown at low density will form less cell-to-cell contacts and therefore will be less likely to communicate via gap junctions. When subjected to photobleaching the fluorescence recovery will be impaired when compared to cells in a confluent monolayer. (**b**) An example of a FRAP curve obtained for a cell (confluent monolayer culture) undergoing photobleaching and subsequent fluorescence recovery. Changes in fluorescence intensity over time are shown. After a sharp decline in fluorescence the intensity increases due to gap junctional communication

3. The mobile fraction percentage is then calculated using the equation:

$$\text{mobile fraction percentage} = \left[\frac{\left(F_m - F_0\right)}{\left(F_{\text{initial}} - F_0\right)}\right] \times 100$$

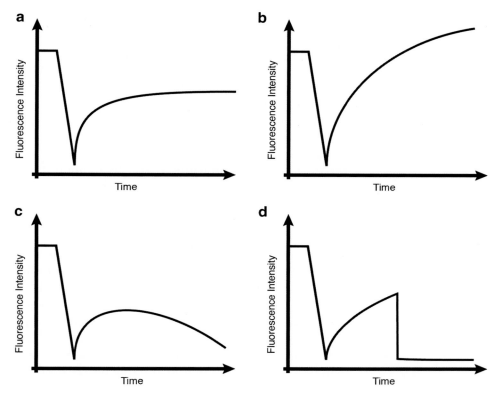

Fig. 3 Types of curves encountered when measuring gap junctional communication using the FRAP method and common problems associated with them. (**a**) Typical recovery curve with a sharp increase in fluorescence intensity after bleaching followed by a plateau. Common issues include (**b**) Recovery without a plateau which indicates that dye is still present in the medium and taken up by the cell during the course of the experiment. (**c**) A decrease in fluorescence intensity over time in the recovery stage that indicates bleaching of the sample during the recovery imaging or the sample going out of focus. (**d**) Initial recovery followed by a sharp decline in fluorescence indicates that the sample has dried out or cell death

4   Notes

1. Cells in low-density cultures form fewer cell-to-cell contacts and therefore are likely to show impaired gap junctional communication when compared to high-density cultures. The low-density cultures can be used as a simple control, i.e., fluorescence recovery should not be seen in single cells not connected to other cells. Other methods include the use of a cell type that does not communicate via gap junctions, such as the HeLa cell line, or a gap junctional inhibitor, e.g., glycyrrhetinic acid.

2. Any adherent cells that communicate via gap junctions can be used for this protocol. Both primary cells and cell lines derived from human and other species are suitable.

3. When comparing different experimental conditions it is important to use cultures with comparable levels of cell confluency.

4. Calcein is a membrane-impermeable hydrolysis product of calcein-AM that is also commonly utilized as a viability indicator. Therefore calcein labeling can also be used to control for the viability of cells subjected to FRAP.

5. It is possible to stain monolayer cultures, 3D cell cultures and small pieces of tissue with calcein-AM. The protocol for staining is the same although it might be necessary to increase the time of incubation with the dye. In addition the samples might require an immobilization step to prevent problems with going out of focus during the experiment. It is also crucial to prevent the samples from drying out during the labeling step as well as at any other step in the protocol.

6. Both photobleaching and fluorescence imaging of calcein make use of an argon-ion laser (488 nm) and a piezomultiplier tube for detection of fluorescence between 500 and 550 nm.

7. To obtain a suitable laser power setting for minimal photobleaching, the average intensity of the bleach region can be monitored over time until a plateau is reached (Fig. 3a). If this does not occur (e.g., Fig. 3c), the laser power during image acquisition may be too high. If lowering the laser power results in a dark or noisy image, the pinhole should be opened until sufficient photons can be detected.

8. In tenocytes, the initial recovery of calcein is rapid. In order to model the recovery as accurately as possible, the time-lapse acquisition must record as much of the initial stages of recovery as possible.

9. A significant portion of the recovery may be missed when the microscope is changing modes between bleaching and time-lapse acquisition modes. Therefore, in order to reduce this changeover time, the zoom can be increased to a value so that the cell is still within the field of view.

10. Scanning is performed unidirectionally with fast scan speed (e.g., 9 on the Zen 2009 platform) and low averaging (e.g., setting of 1 on the Zen 2009 platform) to enable fast recording of the fluorescent recovery. Post-bleaching acquisition should be performed every 5 s for up to 4 min.

11. Gap junction communication can be blocked by incubating the cells with 100 μM 18 β-glycyrrhetinic acid (GA) in DMEM-F12 supplemented with 10 % FBS. The inhibitor should be present during the whole course of the FRAP experiment. GA stock solution can be prepared in dimethyl sulfoxide (DMSO) and stored at −20 °C.

Acknowledgements

Our studies to develop the published methods were supported by Orthopaedic Research UK, Arthritis Research UK, the Oxford NIHR Musculoskeletal Biomedical Research Unit and the Botnar Research Centre Bioimaging Small Research Facility, University of Oxford.

References

1. Reits EA, Neefjes JJ (2001) From fixed to FRAP: measuring protein mobility and activity in living cells. Nat Cell Biol 3:E145–E147

2. Staras K, Mikulincer D, Gitler D (2013) Monitoring and quantifying dynamic physiological processes in live neurons using fluorescence recovery after photobleaching. J Neurochem 126:213–222

3. Wade MH, Trosko JE, Schindler M (1986) A fluorescence photobleaching assay of gap junction-mediated communication between human cells. Science 232:525–528

4. Abbaci M, Barberi-Heyob M, Blondel W et al (2008) Advantages and limitations of commonly used methods to assay the molecular permeability of gap junctional intercellular communication. Biotechniques 45(33–52):56–62

5. Kuzma-Kuzniarska M, Yapp C, Pearson-Jones TW et al (2014) Functional assessment of gap junctions in monolayer and three-dimensional cultures of human tendon cells using fluorescence recovery after photobleaching. J Biomed Opt 19:15001, 1-15001-7

6. Abbaci M, Barberi-Heyob M, Stines JR et al (2007) Gap junctional intercellular communication capacity by gap-FRAP technique: a comparative study. Biotechnol J 2:50–61

7. Waggett AD, Benjamin M, Ralphs JR (2006) Connexin 32 and 43 gap junctions differentially modulate tenocyte response to cyclic mechanical load. Eur J Cell Biol 85:1145–1154

Chapter 13

Tracking Dynamic Gap Junctional Coupling in Live Cells by Local Photoactivation and Fluorescence Imaging

Song Yang and Wen-Hong Li

Abstract

Intercellular communication through gap junction channels is crucial for maintaining cell homeostasis and synchronizing physiological functions of tissues and organs. In this chapter, we present a noninvasive fluorescence imaging assay termed LAMP (local activation of a molecular fluorescent probe) that consists of the following steps: loading cells with a caged and cell permeable coumarin probe (NPE-HCCC2/AM), locally photolyzing the caged coumarin in one or a subpopulation of coupled cells, monitoring cell–cell dye transfer by digital fluorescence microscopy, and post-acquisition analysis to quantify the rate of junction dye transfer using Fick's equation. The LAMP assay can be conveniently carried out in fully intact cells to assess the extent and degree of cell coupling, and is compatible with other fluorophores emitting at different wavelengths to allow multicolor imaging. Moreover, by carrying out multiple photo-activations in a coupled cell pair, LAMP assay can track changes in cell coupling strength between coupled cells, hence providing a powerful method for investigating the regulation of junctional coupling by cellular biochemical changes.

Key words Caged coumarin, Caged dye, Gap junction communication, Infrared-LAMP assay, LAMP assay, NPE-HCCC2/AM, Photoactivation

1 Introduction

Intercellular communication through gap junction channels is known to play important roles in cell homeostasis and synchronization of physiological functions of tissues and organs [1–3]. In vertebrates, junctional coupling is mediated by connexins, a family of proteins that oligomerize to form intercellular gap junction channels [4]. Since malfunction of junctional coupling and mutation of connexins can cause human diseases [5], it is important to study and to understand mechanisms governing intercellular gap junction communication in different biological systems. Methods designed to follow changes in coupling strength or the rates of molecular transfer among coupled cells would certainly facilitate such investigations.

Mathieu Vinken and Scott R. Johnstone (eds.), *Gap Junction Protocols*, Methods in Molecular Biology, vol. 1437,
DOI 10.1007/978-1-4939-3664-9_13, © Springer Science+Business Media New York 2016

Besides electrophysiological approaches, a number of optical methods have been developed to assay gap junction coupling. Essentially all these methods are based on fluorescence imaging of a soluble fluorescent tracer that can diffuse through gap junction channels, i.e., tracers with a molecular weight below 1000. In addition, imaging methods that indirectly assess the extent of cell coupling have also been reported. For example, Ca^{2+} imaging and subsequent correlation analysis of Ca^{2+} activity among individual cells has been used to infer cell coupling in cell populations [6].

This chapter describes two fluorescence imaging assays, LAMP [7] and infrared-LAMP [8], that employ a caged and cell membrane permeable coumarin dye NPE-HCCC2/AM [9] to selectively mark a cell or a subpopulation of cells by using the technique of photoactivation (uncaging) [10]. NPE-HCCC2/AM is the neutral ester of a caged coumarin dye NPE-HCCC2 (Fig. 1). It can diffuse across cell membranes and becomes trapped in the cytosol once the AM ester is hydrolyzed by cellular esterases. Upon photoactivation, either with UV light (LAMP assay) or with two photon laser excitation (infrared-LAMP assay), NPE-HCCC2 is converted to a highly fluorescent coumarin dye HCCC2. HCCC2 is fairly small (molecular weight 449 Da) so it can rapidly diffuse to neighboring cells through gap junction channels. The process of intercellular dye diffusion from the donor cell to the recipient cell is then captured by digital fluorescence microscopy using either wide field, confocal, or two photon imaging. Subsequent analysis of time lapse imaging data provides quantitative information on the kinetics of cell–cell dye transfer [7, 8, 11, 12]. To extent this technique to assay cell coupling in vivo, we have also developed another technique termed Trojan-LAMP and applied it to a model organism, the nematode *C. elegans* [13]. The study revealed that early during embryonic development, the pattern of cell coupling in the developing embryo underwent dramatic remodeling and germ cell precursors were isolated from the somatic cell communication compartment early on.

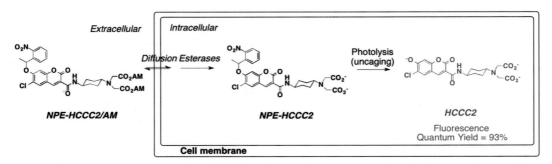

Fig. 1 Structures of a caged and cell permeable coumarin, NPE-HCCC2/AM; its intracellular hydrolysis product, NPE-HCCC2; and its photolyzed product, HCCC2

The LAMP assay possesses several salient features that are desirable for assaying gap junction coupling in live cells. First, because NPE-HCCC2/AM is cell membrane permeable, the method is applicable to fully intact populations including tissues. Second, the uncaging efficiency of NPE-HCCC2 is remarkably high, about two orders of magnitude higher than other caged fluorophores either by UV or two photon photolysis [9]. This extraordinarily high uncaging efficiency makes it possible to photolyze NPE-HCCC2 with a very low dose of UV light or two photon excitation light, hence minimizing phototoxicity or cell damage during uncaging. Together, the above two features makes LAMP assay a truly noninvasive imaging technique. Third, since HCCC2 is a photo-stable and bright fluorescent dye with high fluorescence quantum yield, it yields excellent signal for cell imaging. This in turn produces good signal to background ratio to facilitate quantification of intercellular dye transfer rate. Finally, HCCC2 dye emits blue light that spectrally complements many popular fluorescent sensors emitting at green, yellow or red. This facilitates multicolor imaging to allow imaging cell coupling and other biochemical events in cells concurrently. For instance, we demonstrated that Ca^{2+} influx through store operated Ca^{2+} channel potently inhibited cell coupling by combining the LAMP assay with Fluo-3 calcium imaging [7, 11].

Both the LAMP assay and infrared-LAMP assay have been successfully applied to cultured cell lines [7, 11], freshly isolated primary cells [14], or freshly dissected tissues [8]. We describe herein the LAMP assay and infrared-LAMP assay performed in cultured cells expressing Cx43 and GFP, providing detailed procedures for dye loading, local uncaging and imaging, as well as post-acquisition data analysis and quantification.

2 Materials

2.1 Cell Preparation

1. Cell lines: In addition to primary cells expressing endogenous connexins, cell lines, e.g., Hela cells (*see* **Note 1**) can also be used to study gap junction communication after transfecting these cells with plasmids containing connexin genes (*see* **Note 2**).

2. Plating cells: cells are seeded at about 30% confluence on 3.5 cm imaging dishes containing a glass bottom (MatTek, *see* **Note 3**).

3. Cell transfection: A number of transfection reagents including *Trans*IT-LT1 (Mirus Bio Corporation) can be used to transfect adherent cells. After transfection cultured cells for 24–36 h to allow sufficient expression of connexin proteins (*see* **Note 4**).

4. Cell incubator (37 °C $\pm$ 1 °C, 90% $\pm$ 5% humidity, 5% $\pm$ 1% CO_2).

5. Cell culture medium: DMEM medium containing 10% FBS and 1% penicillin–streptomycin.

6. Trypsin–EDTA and DPBS.

2.2 Dye Loading Solution

1. 10% Pluronic stock solution: dissolve 50 mg pluronic in 0.5 mL anhydrous DMSO to prepare a stock solution. The resulting solution can be stored at 20–25 °C for a couple of months.

2. 2 mM NPE-HCCC2/AM stock solution: Dissolve 1 mg NPE-HCCC2/AM in 670 μL anhydrous DMSO. Store the stock solution at –20 °C or leave it on ice during experiments (*see* **Notes 5** and **6**).

3. 2 mM Fluo-3/AM or calcein/AM stock solutions: Dissolve 1 mg Fluo-3/AM or calcein/AM in anhydrous DMSO (440 μL or 500 μL respectively). Cells are typically loaded with 0.5 mM of calcein/AM, or with 1–2 mM of Fluo-3/AM (*see* **Note 7**).

4. 1× HBS solution: 10× Hank's Balanced Salt Solution (Gibco) is diluted ten times with water. During dilution, Hepes buffer is added to a final concentration of 20 mM, and the pH is adjusted to 7.35 with NaOH or HCl. The solution is then sterilized by filtering through a 0.22 m cellulose filter, aliquoted in 50 mL tubes and stored at 4 °C.

5. DMSO, >99.8%.

6. Pluronic F127, powder.

7. Plasmid DNA for connexins.

8. Mirus transfection reagent.

9. An inverted fluorescence microscope equipped with a field diaphragm, CCD camera, and excitation and emission filter wheels controlled by computer and imaging software. Two photon uncaging was performed on a Zeiss LSM510 imaging system equipped with a Chameleon-XR laser (Coherent) [12].

3 Methods

3.1 Preparation of Dye-Loading Solution

1. Mix 1.2 μL of the pluronic stock solution (10%,) and 1.2 μL of NPE-HCCC2/AM in a 1.5 mL centrifuge tube.

2. Add 0.5 mL of HBS to the mixture and briefly vortex the solution to ensure thorough mixing.

3. If multiple LAMP assays need to be performed on a given day, a larger volume of loading solution can be prepared at once.

4. Once prepared, the loading solution should be stored on ice or at 4 °C and used on the same day (*see* **Notes 8** and **9**).

3.2 Dye Loading

1. Remove the culture medium from the imaging dish by using a disposable Pasteur pipette.

2. Wash the cells twice with 1× HBS solution

3. Remove the HBS solution from the imaging dish and add 0.5–1 mL of the preprepared dye-loading solution to the imaging dish (*see* **Note 10**).

4. Cover the imaging dish with a lid to minimize water evaporation.

5. Incubate cells in the loading solution in the dark for 20–45 min (*see* **Note 11**).

6. Remove the loading solution with a Pasteur pipette and rinse cells once with HBS solution.

7. Add 1.5 mL of HBS solution to the imaging dish.

8. Incubate the cells in the dark for 10 min to allow complete hydrolysis of AM ester and to trap NPE-HCCC2.

3.3 Imaging Assays Using the LAMP Assay

1. The imaging dish containing cells loaded with NPE-HCCC2 is placed on the stage of a microscope.

2. Bring cells into focus by observing cellular fluorescence of a fluorescent marker, e.g., GFP, calcein, or Fluo-3 (excitation 490 ± 10 nm, emission 525 nm ± 10 nm).

3. Once a fluorescent cell pair in close contact is identified, move the stage to center the cell pair in the field of view (Fig. 2a, b).

4. To perform local uncaging in a single cell among a coupled cell pair, place a field diaphragm in the excitation light path to control the diameter of excitation beam.

5. Reduce the field diaphragm to a minimum, such that only a portion of the chosen donor cell is covered by the excitation beam. The recipient cell of the coupled cell pair should be away from the reduced excitation beam (Fig. 2c).

6. Make fine adjustments to position the coupled cell pair with respect to the reduced excitation light beam. The goal is to selectively uncage the donor cell by UV excitation while avoiding photolyzing the recipient cell. Ensure that the cell–cell contact region between the donor cell and the recipient cell stays clear from the reduced uncaging beam (Fig. 2a–c).

7. After adjusting the stage position, open the iris of the field diaphragm and start image acquisition. Acquire two-color image pairs (blue/coumarin channel: Excitation 425 ± 5 nm, Emission 460 nm ± 10 nm; and green/GFP channel: Excitation 490 ± 10 nm, Emission 525 nm ± 10 nm) every 6–10 s for about a minute. We typically set the exposure time for each channel to 50–200 ms. This provides a time course for the baseline cellular fluorescence, which should remain fairly stable.

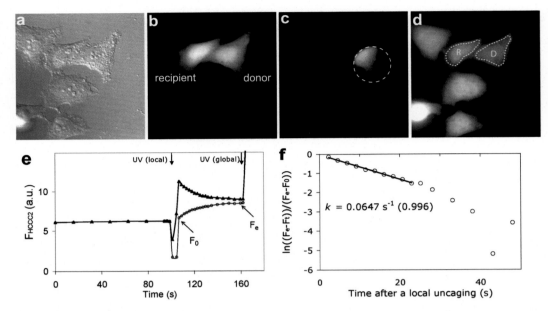

Fig. 2 LAMP assay of gap junctional communication. (**a–c**) DIC (**a**) and GFP images (**b–c**, Excitation 490 ± 10 nm, Emission 525 ± 10 nm) with the iris of the field diaphragm fully open (**b**) or minimized (**c**). The local uncaging area (outlined by the *dashed yellow circle* in **c**) covered a portion of the donor cell and was away from the cell–cell interface. Hela cells were transfected with plasmid containing Cx43-IRES-GFP construct 24 h earlier. Two cells in the field of view were transfected and expressed Cx43 and GFP. (**d**) Coumarin image (Excitation 425 ± 5 nm, Emission 460 ± 10 nm) taken at the end of the experiment after global uncaging showing all cells were loaded with NPE-HCCC2. Two regions of interests (ROIs) representing the donor cell (**d**) and recipient cell (R) are outlined by *dashed yellow circles*. (**e**) Time course of fluorescence intensity of HCCC2 (F_{HCCC2}, arbitrary units) in donor (*filled triangle*) and recipient cells (*open circle*). F_0 denotes F_{HCCC2} immediately after uncaging, and F_e represents F_{HCCC2} when dye transfer reaches equilibrium. Intensity values between F_0 and F_e are F_{HCCC2} at different time points of dye transfer (F_t). (**f**) Quantification of cell–cell dye transfer kinetics by fitting the first half of the dye transfer data from **e**. The slope of the fitted line gave the rate of dye transfer. Adapted from Ref. [12]

8. Prior to local uncaging of the donor cell, increase the image acquisition frequency to about one image pair every 2 s. Reduce the iris of the field diaphragm to a minimum to illuminate only a portion of the donor cell as you set up earlier (Fig. 2c). Viewing from the green channel, reconfirm that the reduced light beam still only targets the donor cell.

9. With the iris of the field diaphragm closed, the donor cell is locally uncaged with a short pulse of UV light (360 ± 20 nm) lasting for 0.1–1 s (*see* **Note 12**).

10. Open the iris of the field diaphragm immediately to continue image acquisition.

11. Depending on the cell coupling strength or the rate of dye transfer, adjust the acquisition frequency by taking a dual-color image pair every 2–10 s.

12. Continue monitoring NPE-HCCC2 transfer from the donor cell to the recipient cell until the equilibrium is reached. At this time the cellular coumarin fluorescence intensities in both the donor and recipient cells come close to each other and remain fairly stable over time (Fig. 2e).

13. Additional uncaging episodes can be performed by repeating **steps 8–12**.

14. The duration of UV exposure should be increased in subsequent uncaging to compensate for the consumption of caged coumarin from the previous photolysis.

15. Before ending the experiment, perform a global uncaging with the field diaphragm fully open to photolyze caged coumarin in all cells. Cells in the field of view should display intense blue coumarin fluorescence (*see* **Note 13**).

3.4 Imaging Assays Using the Two Photon Uncaging and Imaging (Infrared-LAMP Assay)

1. The imaging dish containing cells loaded with NPE-HCCC2 is placed on the stage of a microscope equipped with a two-photon laser (e.g., Zeiss LSM510, or LSM 780 or other equivalent laser scanning imaging systems).

2. Bring cells into focus by observing cellular fluorescence using a fluorescent marker (e.g., GFP, calcein, or Fluo-3) and confocal imaging (Excitation 488 nm laser, Emission 525 nm ± 20 nm, *see* **Note 14**).

3. Similar to the LAMP assay, move the stage to position the coupled cell pair in the center or the field of view.

4. Under 488 nm laser excitation, acquire a z-stack of confocal GFP images.

5. After choosing a donor cell, adjust the focus drive to a z-plane across the middle height of the chosen donor cell. Take a single confocal image at this height.

6. Using LSM510 imaging software, under the "Bleach Control" module specify two photon uncaging laser wavelength (normally between 730 and 760 nm), laser power (~10 mW), uncaging repetitions (20–40) and define the two-photon uncaging area from the confocal image taken above.

7. Use "Define Region" function to draw a circle along the cell periphery of the donor cell (*see* **Note 15**).

8. Start taking time lapse two photon images of cell coumarin fluorescence by exciting the cells at 820 or 830 nm. Take an image every 6–10 s for about a minute. This provides a time course for the baseline cellular fluorescence, which should remain fairly stable.

9. Initiate two-photon uncaging at the predefined donor cell by activating the "Bleach" button. The uncaging laser (760 nm or

below) scans the defined uncaging area repeatedly 20–40 times in a total of a few seconds.

10. The laser excitation wavelength is changed from the uncaging wavelength (760 nm or shorter) to the imaging wavelength (820 or 830 nm) as soon as photo uncaging is finished (*see* **Note 16**).

11. Continue two photon imaging of coumarin fluorescence to monitor dye transfer from the donor to the recipient cell (*see* **Note 17**).

12. To follow dye diffusion in 3D in dissected tissues or in organo-typic cultures, set up a z-stack to center the donor cell along the z-axis.

13. Sample 20–40 z-slices spaced ~2 µm apart either below or above the donor cell. This provides dynamic information of dye transfer from the donor cell to neighboring recipient cells in 3D.

14. Since NPE-HCCC2 can be loaded into cells to fairly high concentrations, additional episodes of dye transfer between coupled cells can be generated by repeating two-photon uncaging in a cell of a coupled cell pair.

15. Between each uncaging episode, cells can be treated with pharmacological or biochemical agents to alter the coupling strength. Their effects on gap junction coupling can then be assessed by comparing the rates of dye transfer before and after drug treatment [7, 12].

3.5 Data Analysis of Dye Transfer Between a Coupled Cell Pair

1. Fick's equation is applied to quantify rates of intercellular coumarin dye transfer. Fick's equation describes kinetics of molecular movement between two compartments separated by a membrane: $\ln[(Fe-Ft)/((Fe-F_0)] = -kt$, where Fe, F_0, and Ft are cellular NPE-HCCC2 signal at equilibrium, zero time and time t, respectively.

2. To quantify cellular NPE-HCCC2 signal, draw separate regions of interest (ROI) above the bulk cytoplasm of the donor cell and the recipient cell to analyze cellular NPE-HCCC2 fluorescence intensity (Fig. 2d). Many imaging software, including OpenLab and ImageJ, can be used to perform such post-acquisition analysis.

3. Plot the measured fluorescence intensities of donor and recipient cells against time. Determine F_0 and Fe from the time course of NPE-HCCC2 signal (Fig. 2e).

4. Replot the data according to Fick's equation (Fig. 2f). Fit the data using the linear equation to obtain the kinetic constant of dye transfer, k (in units of sec^{-1}) (*see* **Note 18**).

4   Notes

1. Primary cells frequently express more than one type of connexins. To study a specific connexin protein, cultured cell lines with very low or undetectable connexin expression can be used, once they are infected with a connexin gene of interest. Cultured cell lines also offer the convenience of easy accessibility and availability.

2. Many plasmids containing connexin genes are available from the open plasmid repository Addgene (https://www.addgene.org/). We have applied LAMP assay in Hela cells infected with plasmids containing Cx43, Cx43-eGFP, Cx43-IRES-eGFP, or Cx26-IRES-DsRed [11].

3. Adhesion of cells to the glass surface usually takes at least 6 h. At low seeding confluence, it is easy to form small cell clusters containing just a few cells (doublets).

4. In this protocol, we used Hela cells expressing Cx43-IRES-eGFP.

5. For long-term storage, make 20–50 μL aliquots of the stock solution and wrap them with aluminum foil to minimize light exposure.

6. NPE-HCCC2/AM is available from VitalQuan LLC or can be synthesized as previously described [9].

7. Calcein normally gives rise to fairly bright fluorescence in cells, so less calcein is needed for cell loading.

8. To aid cell visualization and to assist focusing during fluorescence imaging, a membrane-permeable fluorescent dye such as Fluo-3/AM or calcein/AM can be added to the loading solution. These dyes emit green fluorescence so they will not interfere with imaging blue coumarin signal.

9. Pluronic is a mild detergent that helps to solubilize hydrophobic compounds in aqueous solutions. It greatly improves the cell loading efficiency of AM esters or lipophilic fluorescent probes. The pluronic stock solution may turn cloudy during storage but it can be clarified by warming it at 37 °C or higher temperature for several minutes.

10. Make sure that cells on the center glass coverslip are covered by the dye-loading solution. If necessary, cells on the plastic surface and away from the center glass can be wiped off with a piece of folded Kimwipes. The dried plastic surface will help to retain the loading solution to the center glass.

11. When the confluence of cells is high (>60%), longer incubation may be required to load sufficient amount of caged coumarin into cells. Alternatively, higher concentration (>2 mM)

and/or larger volume of stock solution of NPE-HCCC2/AM can be used to load more cells.

12. The optimal duration of UV photolysis varies with the UV light intensity, or the amount of caged coumarin loaded into cells. Ideally, UV uncaging should produce a sizable increase of coumarin fluorescence that is at least three times above the baseline level. Excess photolysis may generate too strong a coumarin signal that saturates the detector, which should be avoided.

13. If coumarin fluorescence appears to be too weak or too strong after complete photolysis of NPE-HCCC2, the amount of NPE-HCCC2 loaded into cells needs to be adjusted accordingly by, for example, changing the concentration of NPE-HCCC2/AM in the loading solution.

14. Fluo-3 fluorescence tends to be fairly dim at resting cellular Ca^{2+} level. Signal of calcein or GFP is much stronger.

15. The defined uncaging area should be restricted within the donor cell to avoid photolyzing neighboring cells.

16. The old LSM510 imaging software usually takes about 30 s to tune the laser before showing that the laser is mode-locked at the new wavelength. Newer versions of imaging system (LSM780 or more recent ones) tune the laser much faster.

17. The time interval of image acquisition is typically set to be 10–20 s since cell–cell dye transfer usually taking several minutes to reach equilibrium.

18. We normally only fit the first half of dye transfer data because changes in fluorescence intensity over time become smaller and noisier as the dye transfer approaches equilibrium (Fig. 2f).

Acknowledgements

This work was supported by a grant award from NIH (R01 GM077593).

References

1. Goodenough DA, Paul DL (2009) Gap junctions. Cold Spring Harb Perspect Biol 1:a002576

2. Mathias RT, White TW, Gong X (2010) Lens gap junctions in growth, differentiation, and homeostasis. Physiol Rev 90:179–206

3. Pereda AE (2014) Electrical synapses and their functional interactions with chemical synapses. Nat Rev Neurosci 15:250–263

4. Harris AL (2001) Emerging issues of connexin channels: biophysics fills the gap. Q Rev Biophys 34:325–472

5. Kelly JJ, Simek J, Laird DW (2015) Mechanisms linking connexin mutations to human diseases. Cell Tissue Res 360:701–721

6. Hodson DJ, Mitchell RK, Bellomo EA et al (2013) Lipotoxicity disrupts incretin-regulated human beta cell connectivity. J Clin Invest 123:4182–4194

7. Dakin K, Zhao YR, Li WH (2005) LAMP, a new imaging assay of gap junctional communication unveils that Ca^{2+} influx inhibits cell coupling. Nat Methods 2:55–62

8. Dakin K, Li WH (2006) Infrared-LAMP: two-photon uncaging and imaging of gap junctional communication in three dimensions. Nat Methods 3:959

9. Zhao Y, Zheng Q, Dakin K et al (2004) New caged coumarin fluorophores with extraordinary uncaging cross sections suitable for biological imaging applications. J Am Chem Soc 126:4653–4663

10. Li WH, Zheng G (2012) Photoactivatable fluorophores and techniques for biological imaging applications. Photochem Photobiol Sci 11:460–471

11. Dakin K, Li WH (2006) Local Ca^{2+} rise near store operated Ca^{2+} channels inhibits cell coupling during capacitative Ca^{2+} influx. Cell Commun Adhes 13:29–39

12. Yang S, Li WH (2009) Assaying dynamic cell-cell junctional communication using noninvasive and quantitative fluorescence imaging techniques: LAMP and infrared-LAMP. Nat Protoc 4:94–101

13. Guo YM, Chen S, Shetty P et al (2008) Imaging dynamic cell-cell junctional coupling in vivo using Trojan-LAMP. Nat Methods 5:835–841

14. Schumacher JA, Hsieh YW, Chen S et al (2012) Intercellular calcium signaling in a gap junction-coupled cell network establishes asymmetric neuronal fates in C. elegans. Development 139:4191–4201

Chapter 14

A Functional Assay to Assess Connexin 43-Mediated Cell-to-Cell Communication of Second Messengers in Cultured Bone Cells

Joseph P. Stains and Roberto Civitelli

Abstract

Cell-to-cell transfer of small molecules is a fundamental way by which multicellular organisms coordinate function. Recent work has highlighted the complexity of biologic responses downstream of gap junctions. As the connexin-regulated effectors are coming into focus, there is a need to develop functional assays that allow specific testing of biologically relevant second messengers. Here, we describe a modification of the classic gap junction parachute assay to assess biologically relevant molecules passed through gap junctions.

Key words Connexin 43, Cell-to-cell communication, Gap junctions, Luciferase reporter assay, Osteoblast, Transient transfection, Parachute assay

1   Introduction

Cell-to-cell transfer of small molecules through gap junctions regulates diverse biologic processes [1, 2]. In bone, the intercellular communication between gap junction-coupled osteoblast lineage cells is important to tissue function. Recently, a large number of studies have implicated connexin 43 (Cx43)-containing gap junctions in a complex array of biologic responses that influence bone quality in vivo [3–6]. These responses include anabolic and catabolic effects on bone, depending on the cue initiating the cellular response such as mechanical load, disuse, or hormonal challenge. As the biologic effects that occur downstream of messages communicated by gap junctions expand, it becomes imperative to be able to develop assays to define the biologically relevant second messengers that are passed between cells.

We have developed a method that allows the assessment of Cx43-communicated signals by modifying a classic assay for probing gap junction function known as the parachute assay. As originally described [7], the parachute assay takes advantage of

Mathieu Vinken and Scott R. Johnstone (eds.), *Gap Junction Protocols*, Methods in Molecular Biology, vol. 1437,
DOI 10.1007/978-1-4939-3664-9_14, © Springer Science+Business Media New York 2016

a population of donor cells that are loaded with a gap junction-permeable low molecular weight fluorescent dye, such as calcium green AM ester. The donor cells are then seeded onto a confluent monolayer of unlabeled cells, termed the acceptor cells. Gap junctional communication is determined by the diffusion of the low molecular weight fluorophore from the donor cell to the acceptor cell. Later modifications included the use of a non-gap junction transferred dye, such as DiI, to discriminate the donor and acceptor cell populations [8].

The fundamental concept of our assay is the same. We specifically activate signaling in the donor cell population, such as by expression of a constitutively active construct, and introduce a reporter to that signaling event into the acceptor cell population, such as a pathway specific luciferase reporter. Subsequently, we seed the donor cells onto the acceptor cells. The degree of gap junctional communication of the siganling event is measured by reporter activation. Thus, this method assesses not just the passive movement of a tracer molecule from cell to cell, but the functional consequences of intercellular communication. The gap junction dependence of the communication of this signal is verified by increasing or decreasing connexin levels, culturing donor and acceptor cells on transwell chambers and/or by the use of gap junction inhibition.

In the protocol presented below, we transfect the donor cell population with a constitutively active fibroblast growth factor receptor 1 (caFGFR1; myr-FGFR-TDII) [9]. FGFR1 is known to activate phospholipase C gamma 1, which in turn leads to second messengers accumulation [9, 10]. The acceptor cell population is transfected with a Runx2-luciferase reporter construct. We have previously shown that Cx43 amplifies FGF2-dependent signaling to increase the activity of the transcription factor Runx2 [11, 12]. Further, we have shown this involves the inositol pyrophosphate second messenger system [13]. Obviously, this system can be adapted to numerous second messenger-generating effectors in the donor cell and diverse readouts, such as signal pathway specific luciferase reporters and fluorophore activation, in the acceptor cell population. Lastly, this system can be easily adapted to other cell types and other gap junctions. Indeed, we have recently used a similar approach to show the delivery of small interfering RNA between mesenchymal stem cells and synovial fibroblasts in culture [14].

2 Materials

For all tissue culture procedures and reagents used with live cells, aseptic technique and sterile solutions are required. All solutions should be made using ultrapure water. Chemicals should be of molecular biology grade or ACS grade, as available.

2.1 Cell Culture and Transfection

1. MC3T3-E1 clone 4 cells (ATCC, Manassas, USA) (*see* **Notes 1** and **2**).

2. P100 tissue culture dish and 24-well multiwell tissue culture plates.

3. Complete tissue culture medium: α Minimum Essential Medium, 10% fetal bovine serum, 1% penicillin–streptomycin.

4. Calcium magnesium-free Hank's balance salt solution (HBSS): 138 mM sodium chloride, 5 mM potassium chloride, 0.44 mM potassium phosphate monobasic, 0.003 mM sodium phosphate dibasic, 4 mM sodium bicarbonate, 5.6 mM glucose. Sterile filter and store at 4 °C.

5. Tissue culture grade 0.25% trypsin–ethylenediaminetetraacetic acid (EDTA) solution (Thermo Fisher Scientific, USA). Store at −20 °C.

6. Transwell chambers, 5 μm pore size (Corning Life Sciences, USA).

7. Plasmid DNA (*see* **Notes 3** and **4**): we use a pSFFV-neo (i.e., empty vector control for Cx43 overexpression construct), pSFFV-Cx43 (i.e., Cx43 overexpression), constitutively active FGFR1 (i.e., myr-FGFR-TDII, provided by Dr. Daniel Donoghue, University of California San Diego), pcDNA3 (i.e., empty vector control for FGFR1), Runx2 luciferase reporter (i.e., p6xOSE2-Luciferase, provided by Gerard Karsenty, Columbia University) and pRL-TK Renilla luciferase (Promega, USA).

8. Transfection reagents: Jet Prime transfection reagent (Polypus Transfection, France) (*see* **Note 5**).

9. Sterile 1.7 mL eppendorf tubes.

2.2 Luciferase Reporter Assay

1. Dual-injector Centro LB960 Luminometer (Berthold Technologies, USA).

2. HBSS.

3. Luciferase lysis buffer: 25 mM tris base. Adjust pH to 7.8 with phosphoric acid. Add 2 mM dithiothreitol, 2 mM cyclohexylenedinitrilotetraacetic acid (CDTA), 10% glycerol, 1% Triton X-100. Sterile filter and store (*see* **Note 6**).

4. Firefly luciferase assay buffer: 20 mM Tricine, 1 mM magnesium carbonate hydroxide pentahydrate, 2.7 mM magnesium sulfate. Adjust pH to 7.8 with sodium hydroxide. Add 0.01 mM EDTA, 32.4 mM dithiothreitol, 0.063 mM adenosine 5′-phosphate. Sterile filter. Add 0.05 mM D-luciferin and 0.043 mM coenzyme A, lithium salt. Aliquot 10 mL *per* tube and store at −80 °C (*see* **Note 7**).

5. Renilla luciferase assay buffer 1: 1.1 M sodium chloride, 2.2 mM EDTA, 0.22 M potassium phosphate monobasic, 0.575 mM potassium phosphate dibasic, 0.44 mg *per* mL bovine serum albumin, 1.3 mM sodium azide. Adjust pH to 5.0 with hydrochloric acid. Store at 4 °C.

6. 1000× coelenterazine solution: Prepare a 1 M hydrochloric acid solution in methanol. Dissolve 0.00143 mM coelenterazine in the acid–methanol solution. Store at −80 °C in small aliquots.

7. Renilla luciferase working buffer: Dilute coelenterazine solution 1000 times into Renilla luciferase assay buffer 1 just *prior* to use (*see* **Note 8**).

8. 96-well opaque assay plate.

3 Methods

3.1 Cell Culture and Transfection

1. MC3T3 clone 4 osteoblasts cell lines are maintained in complete tissue culture media at 37 °C and 5 % CO_2 in a tissue culture incubator. Cells are passaged at 1:4 to 1:10 *prior* to reaching confluence.

2. One day *prior* to transfection, donor and acceptor cells are seeded at 60,000 cells *per* cm² into a P100 tissue culture treated plate (*see* **Note 9**). For donor cells, there should be four separate sets of plates, namely (1) pSFFV-neo, pcDNA3; (2) pSFFV-Cx43, pcDNA3; (3) pSFFV-neo, caFGFR1; (4) pSFFV-Cx43, caFGFR1. For acceptor cells, there should be two separate sets of plates, namely (1) pSFFV-neo, p6xOSE2-Luc, pRL-TK and (2) pSFFV-Cx43, p6xOSE2-Luc, pRL-TK.

3. Donor cell preparation: Label 4 sterile 1.7 mL eppendorf tubes, namely (1) pSFFV-neo, pcDNA3; (2) pSFFV-Cx43, pcDNA3; (3) pSFFV-neo, caFGFR1; (4) pSFFV-Cx43, caFGFR1. To each tube add 500 μL JetPrime buffer. Then pipet 8 μg of pSFFV-neo or pSFFV-Cx43 and 4 μg of pcDNA or caFGFR1 to each tube, as appropriate. Vortex the samples for 10 s to mix the reagents. Next, add 48 μL JetPrime reagent to each tube (*see* **Note 10**). Vortex the samples for 10 s and incubate at room temperature for 10 min. In a drop wise fashion, pipet the transfection mix onto the cells in the appropriately labeled plate. Swirl the plate gently to mix and return to the incubator. After 4 h, replace the culture media on the cells with fresh complete tissue culture media. Return to the incubator.

4. Acceptor cell preparation: Label 2 sterile 1.7 mL eppendorf tubes, namely (1) pSFFV-neo, p6xOSE2-Luc, pRL-TK and (2) pSFFV-Cx43, p6xOSE2-Luc, pRL-TK. To each tube add 500 μL JetPrime buffer. Then pipet 8 μg of pSFFV-neo or pSFFV-Cx43 and 4 μg of pOSE2 and 1 μg of pRL-TK plasmid

to each tube, as appropriate Vortex the samples for 10 s. Next, add 52 μL JetPrime reagent to each tube. Vortex the samples for 10 s and incubate at room temperature for 10 min. In a dropwise fashion, pipet the transfection mix onto the cells in the appropriately labeled plate. Swirl the plate gently to mix and return to the incubator. After 4 h, replace the media on the cells with fresh complete tissue culture media. Return to the incubator.

5. Coculture with cell-cell contacts: 48 h posttransfection, wash both the donor and acceptor cell cultures two times with HBSS to remove any residual media. Trypsinize the cells from the tissue culture plates with 1 mL 0.25 % trypsin–EDTA solution at 37 °C for about 5 min until the cells round up. Resuspend the cells in 9 mL of complete tissue culture media, transfer to a 15 mL sterile conical tube and pellet the cells by centrifugation at $500 \times g$ for 10 min. Resuspend the cell pellet in 10 mL complete tissue culture media. For acceptor cells, plate 50,000 cells *per* well into a 24-well multiwell plate. For each group, plate 3–6 replicates. Immediately after seeding the acceptor cells, add the appropriate cocultured donor cells at a density of 150,000 cells per well into the same well (*see* **Note 11**) (Fig. 1). Return the cells to incubator for 16 h (*see* **Note 12**). Then proceed to the luciferase reporter assay.

6. Coculture in transwell chambers: Repeat the plate layout as in the preceding step; however, the acceptor cells get plated in the bottom of the 24-well multiwell plate. Then insert the transwell chamber into the well and seed the donor cells into the insert, precluding direct cell-to-cell contact between the donor and acceptor cells. Return the cells to incubator for 16 h. Then proceed to the luciferase reporter assay.

3.2 Luciferase Reporter Assay

1. Remove the culture media from the cells and rinse twice in 1.0 mL HBSS. For cells cultured in transwell chambers, the donor cells in the transwell inserts can be discarded and the acceptor cells rinsed with HBSS (*see* **Note 13**).

DONOR Cells	pSFFV-neo; pcDNA3	pSFFV-Cx43; pcDNA3	pSFFV-neo; caFGFR1	pSFFV-Cx43; caFGFR1
ACCEPTOR Cells	pSFFV-neo; p6xOSE2-Luc	pSFFV-Cx43; p6xOSE2-Luc	pSFFV-neo; p6xOSE2-Luc	pSFFV-Cx43; p6xOSE2-Luc

Fig. 1 Matrix of the coculture seeding setup for conducting the parachute assay. In this example, a 3:1 ratio of donor–acceptor cells were seeded together in the indicated combinations. Cells of the matrix are labeled with the specific combination of co-transfected plasmids for modulating Cx43 expression as well as constitutively active FGFR1 expression

2. Add 200 μL *per* well luciferase lysis buffer. Incubate at room temperature for 30 min with gentle shaking.

3. Transfer 50 μL of each lysate into the wells of an opaque 96-well assay plate.

4. Pre-load 10 mL of firefly luciferase assay buffer into pump 1 of the Berthold Centro LB960 Luminometer. Likewise, pre-load 10 mL of renilla luciferase working buffer into pump 2.

5. Insert the plate into the luminometer and program it to: (1) dispense 100 μL from pump 1 with a by well measurement operation; (2) delay 2 s to allow mixing; (3) measurement for 20 s; (4) dispense 100 μL from pump 2 with a by well measurement operation; (5) delay 2 s to allow mixing; (6) measurement for 20 s.

6. Relative luciferase activity can be determined by dividing the firefly luciferase activity by the renilla luciferase activity. Next, average the replicates together and graph the data.

7. A gap junction communicated signal will result in the activation of the luciferase reporter only when the donor cell expresses both the constitutively active signal protein and Cx43 and the acceptor cell expresses both Cx43 and the luciferase reporter construct and the cells are cocultured in direct contact (Fig. 2).

4   Notes

1. This assay can be adapted to other cell types. The MC3T3 cell line used here has a modest amount of endogenous Cx43 expression. In this context, overexpression or knockdown of Cx43 levels can impact the degree of cell-to-cell coupling and downstream signaling. We have also used this assay in UMR106 osteoblast-like osteosarcoma cells with little endogenous Cx43 expression and in ROS17/2.8 cells with robust Cx43 expression. In the case of low endogenous Cx43 abundance as in UMR-106 cells, overexpression of Cx43 is required to detect cell-to-cell communication, while small interfering RNA-mediated knockdown is ineffective. The converse is true in ROS17/2.8 cells.

2. MC3T3 clone 4 cells maintain a relatively stable phenotype in culture, but in our hands passage numbers over 20 can sometimes result in phenotype changes and reduced transfection efficiency.

3. High quality plasmid DNA is critical to effective and reproducible transfection results. We routinely use a Hi-Speed Maxi prep kit (Qiagen, USA) to prepare our transfection grade plasmids.

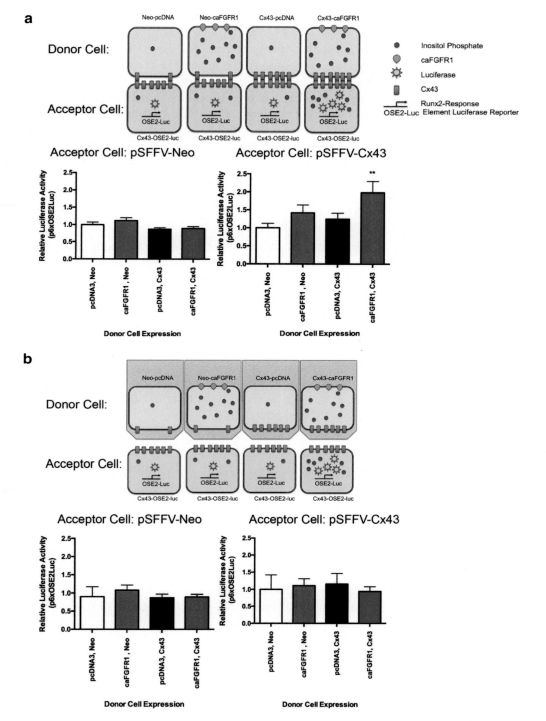

Fig. 2 Representative luciferase reporter data. (**a**) MC3T3 clone 4 cell were cocultured with direct cell-to-cell contact, as indicated in the cartoon. Expression of the caFGFR1 stimulated expression of the p6xOSE2-Luciferase plasmid only in when both the donor and acceptor cells expressed Cx43 containing plasmids (*red bar*). Histograms represent average relative luciferase activity from triplicate samples. Error bars indicated standard deviations. **p-value < 0.01. (**b**) The MC3T3 clone 4 cells were cultured as indicated above, except donor cells were seeded into a transwell chamber without direct cell-to-cell contact with the acceptor cells. In this context, the donor cells were unable to stimulate luciferase activity in the acceptor cells, independent of the Cx43 status

4. When performing overexpression studies, the promoter driving expression of the gene of interest can have a profound impact on function. In our experience, often the robust over-expression from strong promoters, such as CMV and CAG, can have paradoxical effects on signaling, including strong overexpression and knockdown produce the same results. We have encountered this problem in MC3T3 with Cx43 overexpression. Instead, the pSFFV-Cx43 vector, which is driven by a weaker promoter, performs very well in our hands. For different promoters and plasmid constructs, determination of the optimal concentration for the intended biological consequence may be necessary [15].

5. We have used numerous transfection methods in our laboratory to conduct these assays, including Lipofectamine 2000 (Invitrogen), FuGene 6 (Promega), and calcium phosphate co-transfection.

6. Commercial alternatives are available. For dual firefly/renilla luciferase assays, we have successfully used Promega's passive lysis buffer or renilla luciferase assay lysis buffer.

7. Commercial alternatives are available, including the DLR assay reagents from Promega. Our reagents perform comparably to this reagent in head-to-head tests in our lab using these cells and cost considerable less.

8. Our renilla luciferase working reagent is based on a paper by Dyer and colleagues [16].

9. These high plating densities provide our best transfection efficiencies, support cell survival and allow the cells to remain in contact, a necessary condition to study gap junctions. In our hands, osteogenic cells do not proliferate very robustly post-transfection. While these cell densities have produced reproducible data for us in these cell types, optimal cell densities may have to be empirically determined for other cell types.

10. For transfection of MC3T3 clone 4 osteoblasts, we have found that a 4:1 JetPrime–DNA ratio is far superior to the manufacturer's recommendation.

11. We have successfully used donor–acceptor cell ratios that span 1:4 to 4:1. A 3:1 ratio was used in the example provided here. We typically find that when you have a sensitive readout for your acceptor cells, that a higher donor–acceptor cell ratio is beneficial. However, some insensitive readouts, such as the Cx43-dependent small interfering RNA transfer studies we performed [14] require a larger number of acceptor cells to reliably detect changes.

12. We have examined time courses for these effects on luciferase reporter assays, which require signaling, luciferase gene

expression, and luciferase protein synthesis, and found that the minimum amount of time required to detect reproducible effects is 4 h. However, typically coculture for 16–24 h produces a larger effect.

13. Calcium ions are potent inhibitors of luciferase activity. Thus, thorough removal of tissue culture media and washing with calcium-free and magnesium-free HBSS is required for optimal luciferase activity.

Acknowledgement

This work was supported by an NIH grant AR036361 to JPS and NIH grant AR041255 to RC.

References

1. Vinken M (2015) Introduction: connexins, pannexins and their channels as gatekeepers of organ physiology. Cell Mol Life Sci 72:2775–2778

2. Nielsen MS, Axelsen LN, Sorgen PL et al (2012) Gap junctions. Compr Physiol 2:1981–2035

3. Plotkin LI (2014) Connexin43 hemichannels and intracellular signaling in bone cells. Front Physiol 5:131

4. Lloyd SA, Loiselle AE, Zhang Y et al (2014) Shifting paradigms on the role of connexin43 in the skeletal response to mechanical load. J Bone Miner Res 29:275–286

5. Buo AM, Stains JP (2014) Gap junctional regulation of signal transduction in bone cells. FEBS Lett 588:1315–1321

6. Stains JP, Watkins MP, Grimston SK et al (2014) Molecular mechanisms of osteoblast/osteocyte regulation by connexin43. Calcif Tissue Int 94:55–67

7. Ziambaras K, Lecanda F, Steinberg TH et al (1998) Cyclic stretch enhances gap junctional communication between osteoblastic cells. J Bone Miner Res 13:218–228

8. Yellowley CE, Li Z, Zhou Z et al (2000) Functional gap junctions between osteocytic and osteoblastic cells. J Bone Miner Res 15:209–217

9. Hart KC, Robertson SC, Kanemitsu MY et al (2000) Transformation and Stat activation by derivatives of FGFR1, FGFR3, and FGFR4. Oncogene 19:3309–3320

10. Mohammadi M, Honegger AM, Rotin D et al (1991) A tyrosine-phosphorylated carboxy-terminal peptide of the fibroblast growth factor receptor (Flg) is a binding site for the SH2 domain of phospholipase C-gamma 1. Mol Cell Biol 11:5068–5078

11. Lima F, Niger C, Hebert C et al (2009) Connexin43 potentiates osteoblast responsiveness to fibroblast growth factor 2 via a protein kinase C-delta/Runx2-dependent mechanism. Mol Biol Cell 20:2697–2708

12. Niger C, Buo AM, Hebert C et al (2012) ERK acts in parallel to PKCdelta to mediate the connexin43-dependent potentiation of Runx2 activity by FGF2 in MC3T3 osteoblasts. Am J Physiol Cell Physiol 302:C1035–C1044

13. Niger C, Luciotti MA, Buo AM et al (2013) The regulation of runt-related transcription factor 2 by fibroblast growth factor-2 and connexin43 requires the inositol polyphosphate/protein kinase Cdelta cascade. J Bone Miner Res 28:1468–1477

14. Liu S, Niger C, Koh EY et al (2015) Connexin43 mediated delivery of ADAMTS5 targeting siRNAs from mesenchymal stem cells to synovial fibroblasts. PLoS One 10:e0129999

15. Gibson TJ, Seiler M, Veitia RA (2013) The transience of transient overexpression. Nat Methods 10:715–721

16. Dyer BW, Ferrer FA, Klinedinst DK et al (2000) A noncommercial dual luciferase enzyme assay system for reporter gene analysis. Anal Biochem 282:158–161

Calcium Wave Propagation Triggered by Local Mechanical Stimulation as a Method for Studying Gap Junctions and Hemichannels

Jegan Iyyathurai, Bernard Himpens, Geert Bultynck, and Catheleyne D'hondt

Abstract

Intercellular communication is essential for the coordination and synchronization of cellular processes. Gap junction channels play an important role to communicate between cells and organs, including the brain, lung, liver, lens, retina, and heart. Gap junctions enable a direct route for ions like calcium and potassium, and low molecular weight compounds, such as inositol 1,4,5-trisphosphate, cyclic adenosine monophosphate, and various kinds of metabolites to pass between cells. Intercellular calcium wave propagation evoked by a local mechanical stimulus is one of the gap junction assays to study intercellular communication. In experimental settings, an intercellular calcium wave can be elicited by applying a mechanical stimulus to a single cell. Here, we describe the use of monolayers of primary bovine corneal endothelial cells as a model to study intercellular communication. Calcium wave propagation was assayed by imaging fluorescent calcium in bovine corneal endothelial cells loaded with a fluorescent calcium dye using a confocal microscope. Spatial changes in intercellular calcium concentration following mechanical stimulation were measured in the mechanical stimulated cell and in the neighboring cells. The active area (i.e., total surface area of responsive cells) of a calcium wave can be measured and used for studying the function and regulation of gap junction channels as well as hemichannels in a variety of cell systems.

Key words Bovine corneal endothelial cells, Calcium imaging, Calcium wave propagation, Connexin 43, Gap junction assay, Intercellular communication, Intracellular calcium, Mechanical stimulation

1 Introduction

Intercellular communication is essential for tissue homeostasis, control of cell proliferation and synchronization of response to extracellular stresses, thereby coordinating the physiological process between or within a variety of organs and tissues, including the brain, lung, liver, lens, retina, and heart [1]. Connexin proteins have been shown to serve as crucial intercellular communication channels in a variety of cell systems and tissues. In vertebrates, 20 different connexin isoforms are expressed [2]. These connexin

Mathieu Vinken and Scott R. Johnstone (eds.), *Gap Junction Protocols*, Methods in Molecular Biology, vol. 1437,
DOI 10.1007/978-1-4939-3664-9_15, © Springer Science+Business Media New York 2016

isoforms are members of the highly conserved multigenic family of transmembrane proteins, serving as the building blocks for both gap junction and hemichannels. Connexin nomenclature is based on predicted molecular weight of the isoform and is based on sequence similarity and length of the cytoplasmic domain of the connexins, thereby classifying them into α, β, and γ subgroups [3]. As such, six connexins, radially arranged around a central pore, form a connexon. Head-to-head docking of two connexons, also called hemichannels, of adjacent cells results in the formation of a gap junction channel. A plaque of proteinaceous gap junction channels, interconnecting the cytoplasm of adjacent cells forms a gap junction. Gap junctions communicate directly between cells via the diffusion of calcium (Ca^{2+}) or inositol 1,4,5-trisphosphate through gap junctions that couple adjacent cells causing release of Ca^{2+} from intracellular stores. In contrast, hemichannels can communicate via the release of diffusible extracellular messengers, like adenosine triphosphate that can cause a Ca^{2+} transient in neighboring cells via Ca^{2+} influx or via Ca^{2+} release from intracellular stores (Fig. 1). A more detailed discussion on the mechanisms underlying the initiation and occurrence of intercellular Ca^{2+} waves and their physiological relevance is provided elsewhere [4].

A number of techniques is used to study gap junctional communication including microinjection [5], scrape loading [6], fluorescence recovery after photobleaching [7], preloading assay [8], local activation of a molecular fluorescent probe [9], electroporation [10], and dual whole-cell patch clamp [11] and mechanical stimulation [12]. Each gap junction assay has its advantages and limitations [13].

Here, we describe the method of studying intercellular communication by investigating Ca^{2+} wave propagation elicited by mechanical stimulation of a single cell. This technique provides a tool to quantify the spread of the Ca^{2+} wave over time in cell line models and primary cell systems and to compare different cell treatments quantitatively. The Ca^{2+} wave propagation is assayed by intracellular Ca^{2+} imaging. This is done by loading the bovine corneal endothelial cells (BCEC) with the Ca^{2+}-sensitive dye Fluo-4 AM to monitor cytoplasmic Ca^{2+} concentration. The fluorescence intensity of Fluo4 is a quantitative readout for cytoplasmic Ca^{2+} concentration. Fluo4 is excited at 488 nm and its emission is recorded at 530 nm. A neutral density filter is used to minimize photobleaching.

Mechanical stimulation of a single cell consists of an acute short-lasting deformation of the cell by briefly touching less than 1% of the cell membrane with a glass micropipette (i.e., tip diameter < 1 μm) coupled to a piezoelectric crystal nanopositioner, mounted on a micro-manipulator. In BCEC, mechanical stimulation results in a rapid initial Ca^{2+} rise that originates at the point of stimulation and spreads throughout the mechanically stimulated cell,

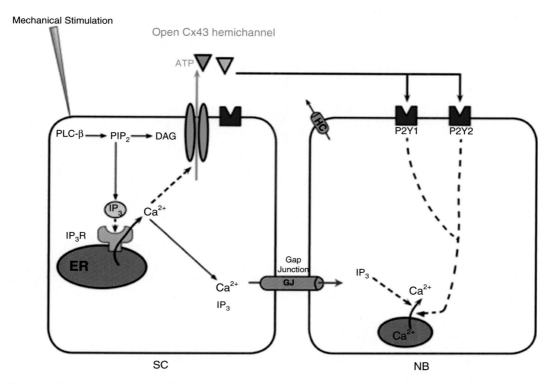

Fig. 1 A schematic model for Ca^{2+} wave propagation in bovine corneal endothelial cells (BCEC). In normal BCEC, it is hypothesized that mechanical stimulation leads to a moderate rise in cytosolic Ca^{2+} concentration via inositol 1,4,5-trisphosphate (IP$_3$)-dependent signaling mechanisms, which leads to the opening of Cx43 hemichannels and the flux of adenosine triphosphate (ATP) from the cytosol into the extracellular environment. This allows the propagation of Ca^{2+} from the "stimulated cell" (SC) to neighboring (NB) cells via activation of purinergic receptors and downstream IP$_3$-induced Ca^{2+} signaling. This figure and its legend have been taken and adapted from D'hondt C, Iyyathurai J, Himpens B, Leybaert L, Bultynck G. (2014) Cx43-hemichannel function and regulation in physiology and pathophysiology: insights from the bovine corneal endothelial cell system and beyond. Front Physiol. 5:348. doi: 10.3389/fphys.2014.00348. eCollection 2014

slowly diminishing to the baseline level. Subsequently, the intracellular Ca^{2+} wave propagates to the surrounding neighboring cells as an intercellular Ca^{2+} wave, upon reaching the cell boundaries. The mechanotransduction-induced Ca^{2+} increase in the mechanically stimulated cell has not yet been fully elucidated, but could be attributed to Ca^{2+} influx and/or to Ca^{2+} release in response to local inositol 1,4,5-trisphosphate production in the mechanical stimulated cells (Fig. 1).

In BCEC, intercellular Ca^{2+} wave propagation is mainly driven by connexin 43 (Cx43)-based hemichannels mediating the release of adenosine triphosphate in the extracellular environment and only a minor part is driven by gap junctional coupling [14, 15]. Using a combination of genetic tools, like small interfering RNA against Cx43, peptide tools that inhibit Cx43 gap junctions and/or hemichannels and adenosine triphosphate-degrading enzymes,

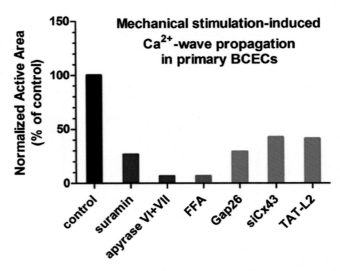

Fig. 2 A graph depicting the characteristics of intercellular communication in BCEC, based on mechanical stimulation-induced Ca^{2+}-wave propagation data (active area) [14–16]. Data were further normalized to their respective controls set at 100 %. The graph is intended to indicate the relevance of adenosine triphosphate (ATP) release (*blue bars*), hemichannels (*red bar*) and Cx43-based hemichannels (*green bars*). In general, the data indicate that in BCEC mechanical stimulation-induced Ca^{2+} wave propagation is almost completely driven by release of ATP into the extracellular environment (i.e., about 90 % inhibition by ATP-degrading enzymes) and that Cx43 hemichannels are a major release pathway for this ATP (i.e., about 60 % inhibition upon Cx43 knockdown or inhibition), although other connexin and/or pannexin isoforms likely contribute to ATP release. Since this graph is intended to provide a general view, readers should access the original research paper for obtaining information about the original mean data and their respective standard error of the mean values. This figure and its legend have been taken from D'hondt C, Iyyathurai J, Himpens B, Leybaert L, Bultynck G. (2014) Cx43-hemichannel function and regulation in physiology and pathophysiology: insights from the bovine corneal endothelial cell system and beyond. Front Physiol. 5:348. doi: 10.3389/fphys.2014.00348. eCollection 2014

it was shown that the active area (i.e., the maximal total surface area of responsive cells) was heavily reduced in cells treated with (1) small interfering RNA against Cx43, (2) TAT-L2, a cell-permeable peptide inhibiting Cx43 hemichannels, while keeping Cx43 gap junctions in an open state or (3) apyrase, an adenosine triphosphate-degrading enzyme (Fig. 2).

Here, we describe in detail the experimental protocol for the measurement of mechanical stimulation-induced Ca^{2+} wave propagation, as performed in BCEC. Besides mechanical stimulation, the properties of intercellular Ca^{2+} waves can also be studied in a quantitative manner through their initiation in a controlled manner upon a local photo-release of caged inositol 1,4,5-trisphosphate, which is described elsewhere [17].

2   Materials

2.1   BCEC Medium Preparation

BCEC growing medium: Dulbecco's modified Eagle's medium (high glucose, no glutamine and no pyruvate) (Thermo Fisher Scientific, Belgium) supplemented with 6.6% GlutaMAX™ (Thermo Fisher Scientific, Belgium), 10% fetal bovine serum (Sigma, Belgium), 1% antibiotic–antimycotic (Thermo Fisher Scientific, Belgium), and 1% Fungizone® antimycotic (Thermo Fisher Scientific, Belgium).

2.2   BCEC Isolation

1. Cell culture dish, 100×20 mm (Sigma, Belgium).
2. Earle's balanced salt solution (EBSS) (Thermo Fisher Scientific, Belgium).
3. Fire-polished hook-shaped glass Pasteur pipette.

2.3   Cell Culture

1. Versene solution (Thermo Fisher Scientific, Belgium).
2. Trypsin–ethylenediaminetetraacetic acid, 0.05% (Thermo Fisher Scientific, Belgium).
3. Laminar air flow cabinet.
4. Hemocytometer.
5. CO_2 incubator (37 °C and 5% CO_2).
6. Culture flasks, 25 and 75 cm² (Sigma, Belgium).
7. Chambered slides, 4.2 cm² (Sigma, Belgium).

2.4   Measurement of Intercellular Calcium Waves Using Mechanical Stimulation

1. Dulbecco's phosphate-buffered saline (DPBS) buffer with Ca^{2+} and Mg^{2+} (Thermo Fisher Scientific, Belgium).
2. DPBS buffer without Ca^{2+} and Mg^{2+} (Thermo Fisher Scientific, Belgium).
3. Fluo-4 AM (Thermo Fisher Scientific, Belgium).
4. LSM510 confocal microscope (Zeiss, Germany).
5. Glass capillaries for nanoliter 2010, 3.5 in. long (World Precision Instruments, UK).
6. Piezoelectric crystal nanopositioner (Piezo Flexure NanoPositioner P-280, operated through E463 amplifier/controller, PI Polytech, Karlsruhe, Germany).
7. Zeitz DMZ-puller (Zeitz Instruments, Germany).

3   Methods

3.1   Procedure of Cell Isolation

1. Isolate the fresh eyes from cow (*see* **Note 1**).
2. Place the eye on a cell culture dish (i.e., 100×20 mm) in a laminar air flow and sterilize by spraying with 70% ethanol.

3. Remove the 70% ethanol from the surface of the cornea by rinsing with EBSS–1% iodine solution.

4. Carefully dissect the cornea from the eye (*see* **Note 2**) and place it in a new cell culture dish (i.e., 100×20 mm), which contains EBSS, with the epithelial cell layer facing upward.

5. Remove any remaining iris tissue still attached to the cornea.

6. Transfer the cornea to another cell culture dish with the endothelial cell layer upward and rinse twice with EBSS.

7. Transfer the cornea to an hourglass and cover with growth medium.

8. Remove the growth medium with a suction pipette, add 300 µL of a trypsin solution to the endothelial layer of the cornea and immediately remove the trypsin solution.

9. Add 300 µL of fresh trypsin solution to the endothelial layer of the cornea and incubate for 30 min at 37 °C with 5% CO_2.

10. Gently scrape the endothelial cells away from the cornea (*see* **Note 3**) and add to culture flasks (i.e., 25 cm²) containing 4 mL of growth medium (*see* **Note 4**).

3.2 Cell Culture

1. The next day, add 5 mL of growth medium to the culture flasks (*see* **Note 5**).

2. Refresh the growth medium every second day until confluency is reached (*see* **Note 6**).

3. Remove the culture medium and wash the cells twice with 5 mL Versene solution.

4. Add 1.5 mL trypsin solution and place it in the incubator for 3–4 min to detach the cells.

5. Thereafter add 12 mL of growth medium to inhibit the trypsin action and pipette the medium three times in and out to disperse the cells, subsequently count the cells using a hemocytometer.

6. Seed the cells in chambered slides with an area of 4.2 cm² with a cell count of 165,000 cells (i.e., cell density is 39,286 *per* cm²) (*see* **Note 7**).

7. Incubate the cells in an incubator at 37 °C with 5% CO_2. Refresh the medium every 2 days until 95% confluency is reached (*see* **Note 8**).

3.3 Measurement of Intercellular Calcium Waves Using Mechanical Stimulation

1. Wash the chambered slide two times with DPBS buffer (*see* **Note 9**).

2. Load the cells with the 500 µL of 10 µM Ca^{2+}-sensitive dye Fluo-4 AM solution (*see* **Note 10**) and incubate for 30 min at 37 °C.

3. Wash the dye three times with DPBS and add 500 µL of DPBS buffer (*see* **Note 11**).

4. The Ca²⁺ wave propagation is assayed by measure spatial changes in intracellular Ca^{2+} concentration following mechanical stimulation using the LSM510 confocal microscope.

5. Use an oil immersion 40× objective and set the confocal microscope using Argon laser by excite at 488 nm (i.e., use beam splitter HFT 488) and collect the fluorescence emission at 530 nm (i.e., using a long pass emission filter LP 505), set the pinhole at minimum.

6. Search the confluent cells and position the glass micropipette at 45° in respect to the chambered slide (*see* **Note 12**).

7. Operate the nanopositioner with voltage between 0.2 and 1.5 V during the mechanical stimulation.

8. Collect and store images.

9. Draw a polygonal region of interest to define the total surface area of responsive cells (i.e., active area, AA) using the software of the confocal microscope.

4   Notes

1. Isolate BCEC from fresh eyes of maximal 18 months old cows, obtained from the slaughterhouse, in order to isolate primary culture of BCEC. Eyes are enucleated at the slaughterhouse within 5 min *postmortem* and preserved in EBSS-1 % iodine solution on ice for transportation to the laboratory, where cell isolation takes place.

2. Use a sterile sharp razor blade to make a deep enough cut through the sclera. Do not press much on eyeball while cutting to avoid ooze out of fluids. Then use sterile forceps to peel the cornea away from the underlying tissue.

3. Use a fire-polished hook-shaped glass Pasteur pipette to scrape the endothelial cells.

4. Repeat the procedure once more.

5. When the BCEC are still not attached to the surface of the flask, put the flask for another 2 days in the incubator.

6. Once attached to the surface of the flask, change the cell culture medium. On average, the cells are confluent in 10 days (i.e., 7–12 days).

7. Transfer the remaining cells to a 75 cm² culture flasks at a density of 6250 per cm² and refresh the culture medium every 2 days with total volume of 20 mL. When confluency is reached, trypsinize the cells and redistribute the cells into chambered slides. This procedure can be repeated twice and cell cultures up to passage 2 can be used for experiments.

8. On average, confluency is reached within 3–4 days.

9. Use DPBS containing Ca^{2+} and Mg^{2+}.

10. In order to open the gap junction channels, prepare 10 μM Fluo-4 AM in DPBS without Ca^{2+} and Mg^{2+}.

11. Now, use DPBS containing Ca^{2+} and Mg^{2+} to close the gap junction channels.

12. Prepare a glass pipette (i.e., tip diameter < 1 μm) using micro-electrode puller and couple it to a piezoelectric crystal nanopositioner, which is operated through an amplifier, mounted on a micro-manipulator.

Disclaimer

Part of this protocol has been previously published as D'hondt C, Himpens B, Bultynck G. (2013) Mechanical stimulation-induced calcium wave propagation in cell monolayers: the example of bovine corneal endothelial cells. J Vis Exp. (77):e50443. doi: 10.3791/50443. A permission from the Journal of Visualized Experiments to reuse parts of the article "D'hondt, C., Himpens, B., Bultynck, G. Mechanical Stimulation-induced Calcium Wave Propagation in Cell Monolayers: The Example of Bovine Corneal Endothelial Cells. J. Vis. Exp. (77), e50443, doi:10.3791/50443 (2013)" for inclusion in Methods in Molecular Biology (Springer) has been obtained.

Acknowledgements

The work has been supported by Concerted Actions of the K.U. Leuven (GOA/09/012), Research Foundation - Flanders (F.W.O.; grant G.0298.11 to GB), Interuniversity Attraction Poles Program (Belgian Science Policy; P7/13 to G.B), and a "Krediet aan Navorser" grant of the FWO (15117.14 N to CDH).

References

1. Kar R, Batra N, Riquelme MA et al (2012) Biological role of connexin intercellular channels and hemichannels. Arch Biochem Biophys 524:2–15

2. Scemes E, Spray DC, Meda P (2009) Connexins, pannexins, innexins: novel roles of "hemichannels". Pflugers Arch 457:1207–1226

3. Sohl G, Willecke K (2004) Gap junctions and the connexin protein family. Cardiovasc Res 62:228–232

4. Leybaert L, Sanderson MJ (2012) Intercellular Ca^{2+} waves: mechanisms and function. Physiol Rev 92:1359–1392

5. Meda P (2000) Probing the function of connexin channels in primary tissues. Methods 20:232–244

6. El-Fouly MH, Trosko JE, Chang CC (1987) Scrape-loading and dye transfer. A rapid and simple technique to study gap junctional intercellular communication. Exp Cell Res 168:422–430

7. Abbaci M, Barberi-Heyob M, Stines JR et al (2007) Gap junctional intercellular communication capacity by gap-FRAP technique: a comparative study. Biotechnol J 2:50–61

8. Goldberg GS, Bechberger JF, Naus CC (1995) A pre-loading method of evaluating gap

junctional communication by fluorescent dye transfer. Biotechniques 18:490–497

9. Dakin K, Zhao Y, Li WH (2005) LAMP, a new imaging assay of gap junctional communication unveils that Ca^{2+} influx inhibits cell coupling. Nat Methods 2:55–62

10. Geletu M, Guy S, Firth K et al (2014) A functional assay for gap junctional examination; electroporation of adherent cells on indium-tin oxide. J Vis Exp 92: e51710

11. Wilders R, Jongsma HJ (1992) Limitations of the dual voltage clamp method in assaying conductance and kinetics of gap junction channels. Biophys J 63:942–953

12. D'hondt C, Himpens B, Bultynck G (2013) Mechanical stimulation-induced calcium wave propagation in cell monolayers: the example of bovine corneal endothelial cells. J Vis Exp 77:e50443

13. Abbaci M, Barberi-Heyob M, Blondel W et al (2008) Advantages and limitations of commonly used methods to assay the molecular permeability of gap junctional intercellular communication. Biotechniques 45(33–52): 56–62

14. Gomes P, Srinivas SP, Vereecke J et al (2005) ATP-dependent paracrine intercellular communication in cultured bovine corneal endothelial cells. Invest Ophthalmol Vis Sci 46:104–113

15. Gomes P, Srinivas SP, Van Driessche W et al (2005) ATP release through connexin hemichannels in corneal endothelial cells. Invest Ophthalmol Vis Sci 46:1208–1218

16. Ponsaerts R, De Vuyst E, Retamal M et al (2010) Intramolecular loop/tail interactions are essential for connexin 43-hemichannel activity. FASEB J 24:4378–4395

17. Decrock E, De Bock M, Wang N et al (2015) Flash photolysis of caged IP_3 to trigger intercellular Ca^{2+} waves. Cold Spring Harb Protoc 3:289–292

Chapter 16

Establishment of the Dual Whole Cell Recording Patch Clamp Configuration for the Measurement of Gap Junction Conductance

Richard D. Veenstra

Abstract

The development of the patch clamp technique has enabled investigators to directly measure gap junction conductance between isolated pairs of small cells with resolution to the single channel level. The dual patch clamp recording technique requires specialized equipment and the acquired skill to reliably establish gigaohm seals and the whole cell recording configuration with high efficiency. This chapter describes the equipment needed and methods required to achieve accurate measurement of macroscopic and single gap junction channel conductances. Inherent limitations with the dual whole cell recording technique and methods to correct for series access resistance errors are defined as well as basic procedures to determine the essential electrical parameters necessary to evaluate the accuracy of gap junction conductance measurements using this approach.

Key words Gap junction conductance, Transjunctional voltage, Patch clamp, Dual whole cell configuration, Series resistance, Membrane potential, Channel conductance, Perforated patch

1 Introduction

The earliest quantitative estimates of the coupling resistance (R_c) between electrically coupled cells were performed by measuring the voltage response in a cell adjacent to a current injected cell or $R_c = \left(\dfrac{V_{m1}}{V_{m2}} - 1 \right) \times R_2 = \dfrac{R_2 \times (V_{m1} - V_{m2})}{R_2}$, where V_{m1} and V_{m2} are the membrane voltage responses in the injected cell and adjacent cell, respectively, and R_2 is the input resistance of the adjacent cell [1]. This arrangement requires separate voltage and current microelectrodes impaled into each cell, four in total, and this technique is only applicable to large cells. This same two-electrode voltage clamp configuration was used to perform the first direct measurement of gap junctional conductance (g_j) by independently voltage clamping two coupled cells and dividing the current change in the

Mathieu Vinken and Scott R. Johnstone (eds.), *Gap Junction Protocols*, Methods in Molecular Biology, vol. 1437, DOI 10.1007/978-1-4939-3664-9_16, © Springer Science+Business Media New York 2016

non-injected cell by the membrane voltage difference between the two cells [2]. This two-electrode two-cell voltage clamp technique led to the first description of the kinetic and steady-state gating of g_j by transjunctional voltage (V_j) gradients [3, 4]. The development of the whole cell patch clamp technique, where one patch electrode serves as both the voltage and current electrode for each cell, permitted the application of the two-cell voltage clamp technique to pairs of small cells (i.e., <20 μm in diameter) and led to the first measurements of single gap junction channel conductances (γ_j) within the next 5 years [5–7]. The cloning of the first two connexins occurred shortly thereafter and the biophysical investigation of connexin-specific gap junctions rapidly evolved with the development of exogenous expression systems of newly cloned connexins using communication-deficient cells or connexin38 (Cx38) antisense injected *Xenopus* oocytes [8–13]. This chapter focuses on the dual whole cell patch clamp methods used to measure g_j and γ_j from paired cells with high input resistances with single channel current resolution.

2   Materials

2.1   Solutions

1. Extracellular (bath) solution: Cultured cells must be thoroughly rinsed with a protein-free physiological saline solution for effective gigaohm (GΩ) seal formation. Typically, this is accomplished by rinsing the dish four to six times with a physiological balanced salt solution (BSS) to remove the serum-containing culture solution, letting the culture dish incubate at room temperature for another 5–10 min to solubilize any remaining serum proteins from the cells and culture dish surface, rinsing the dish one to two times to remove any loose cells, and transferring the culture dish to the microscope stage for patch clamp procedures. The BSS composition is 140 mM NaCl, 1.3 mM KCl, 4.0 mM CsCl, 2.0 mM tetraethyl ammonium chloride (TEACl), 1.0 mM NaH_2PO_4, 1.8 mM $CaCl_2$, 0.8 mM $MgSO_4$, 5.5 mM dextrose, and 10 mM 4-(2-hydroxyethyl)-1-piperazineethanesulfonic acid (HEPES), titrated to pH 7.4 with 1 N NaOH. The low KCl and addition of CsCl and TEACl reduces any plasmalemmal background K^+ currents to near zero and helps to maintain a high cellular input resistance (R_{in}) upon formation of the whole cell patch configuration, thus improving the gap junctional current signal-to-noise ratio. The BSS composition varies among gap junction electrophysiologists but typically consists of 140 mM monovalent salt (i.e., NaCl or CsCl) 1–3 mM divalent cations (i.e., Ca^{2+} or Mg^{2+}), 1 g/L (5.5 mM) d-glucose, 5–10 mM HEPES, titrated to pH 7.2–7.4 with 1 N NaOH or CsOH.

2. Intracellular patch pipette solution (IPS): 140 mM KCl, 1.0 mM $MgCl_2$, 5.0 mM 1,2-bis(2-aminophenoxy)ethane-N,N,N',N'-tetraacetic acid tetrapotassium salt (K_4BAPTA), 3.0 mM $CaCl_2$ (i.e., MAXCHELATOR [14] estimated free Ca^{2+} about 200 nmol/L), and 25 mM HEPES, titrated to pH 7.4 with 1 N KOH.

2.2 Dual Whole Cell Patch Clamp Setup

1. Patch clamp amplifiers: Controlling the V_j gradient between two coupled cells requires two low-noise whole cell patch clamp amplifiers. Examples include the Axopatch 200B or computer-controlled Multipatch 700B amplifiers from Molecular Devices, EPC800 USB or computer-controlled EPC 10/2,3,4 USB amplifiers from Heka Elektronik, or PC-505B from Warner Instruments. Automated patch clamp systems cannot perform the dual whole cell patch clamp voltage clamp method. Each patch clamp amplifier consists of an electronic rack mountable main amplifier and head stage amplifier connected by a flexible shielded cable of 3–6 ft (i.e., 1–2 m) in length and a patch electrode holder designed to fit onto the front of the head stage amplifier. The patch electrode holder will be equipped with a thin silver wire (i.e., 30 AWG, 0.010 in., 0.25 mm diameter), such as AGW1010 from World Precision Instruments (WPI) with the outer half of the wire electroplated with chloride (*see* **Note 1**).

2. XYZ micromanipulators: Stable positioning of each patch electrode on a cell requires a three-axis (XYZ) coarse manual and remote fine control micromanipulator sturdy enough to support the head stage/pipette holder of the selected patch clamp amplifier with submicron resolution and negligible drift. There are multiple choices in the type of fine/coarse control micromanipulators including hydraulic (i.e., preferably water, less viscous drag), piezo-electric, stepper motor, and motorized linear actuator models. Examples include Narishige MHW-3 or MHW-103 water hydraulic, ThorLabs Burleigh 5200/5300/5400 series piezo-electric, Sutter Instruments MPC-200/MPC-285/MPC-225/MPC-265 series or Narishige EMM-3NV stepper motor, Zaber M-LSM linear actuator or Newport Corp. 462XYZ linear stage plus a variety of linear actuator models (*see* **Note 2**). Each two-cell patch clamp setup will require one right-handed and one left-handed version. The angle of the patch electrode should be between 40 and 60° when mounted in the micromanipulator.

3. Inverted microscope: For viewing cells in culture, an inverted light microscope with 10× magnification oculars and 10×, 20×, and 40× long working distance objective lenses is optimal. Patching a cell, forming the GΩ seal, is usually performed under 400× or even 600× magnification. To visualize the

membrane, either phase contrast or Nomarski differential interference contrast (DIC) imaging is necessary (*see* **Note 3**). All major scientific microscope manufacturers offer inverted light microscopes suitable for patch clamping. Examples include the Olympus IX-73 or motorized IX-83, Nikon Eclipse Ti, Leica DMi8, or Zeiss Axio Vert.A1 models.

4. Anti-vibration table: Stable patch clamp recordings require a mechanical vibration-free environment. This requires that the inverted light microscope and XYZ micromanipulators be positioned on an anti-vibration table. Examples include the Technical Manufacturing Corporation (TMC) Micro-G 63-500 series, Newport Integrity series, Kinetic Systems MK26 or MK52 series, ThorLabs Active-Air series, MinusK Microscope vibration isolation, Electron Microscopy Solutions (EMS) AMH series anti-vibration table workstations.

5. Faraday cage: In addition to vibration isolation, isolation of external electrical noise, such as 50–60 Hz AC power sources, is critical since something as minor as room lights may transmit alternating current signals in excess of the biological signal you are attempting to record onto the patch clamp recording chamber. Faraday cages are typically constructed of conductive stainless steel or copper mesh screens or MuMetal (Magnetic Shield Corporation) and cover the entire anti-vibration table top or minimally the microscope stage and micromanipulators. Some anti-vibration table manufacturers, including TMC, Newport, Kinetic Systems, MinusK, ThorLabs, sell Faraday cages for their anti-vibration tables as accessories or one may custom-build one using metal screen wire, copper mesh, or MuMetal to fit a specific patch clamp system.

6. Patch pipette puller: One cannot attain a live whole cell patch clamp recording without preparing fresh patch electrodes on the day of use. The most popular model of patch pipette puller in use is the Sutter Instruments P-97 Flaming/Brown Micropipette puller named after authors of method for its design [15]. Sutter Instruments also markets P-1000 Next Generation and P-2000 laser-based micropipette pullers. Key factors in the development of a suitable patch pipette for whole cell electrophysiological recordings include the choice of capillary glass and program design to pull a patch electrode with an appropriate geometry for whole cell recording (i.e., low access resistance). Typically, a 1.5 mm outside diameter (OD) glass capillary with inside diameters (ID) of 0.75–1.1 mm, 10 cm in length, without a filling filament are used. References for the fabrication of patch pipettes include Rae and Levis [16] and the Sutter Instrument P-97 pipette cookbook [17]. We use a 1.5/0.84 OD/ID borosilicate glass from WPI without fire-polishing or coating with Sylgard® 182 after preparing the patch pipette.

7. Data Acquisition hardware/software: Analog signals of the whole cell voltage clamp currents are recorded in real-time via analog-to-digital (A/D) sampling and computer storage for off-line analysis. Molecular Devices markets its own data Digidata 1550 digitizer (A/D converter) and pClamp10 and Clampfit10 software for data acquisition and analyses using the Axopatch 200B or 700B patch clamp amplifiers. Heka Elektronik similarly markets InstruTECH LIH 8 + 8 and ITC-18 data acquisition interfaces and Patchmaster and Fitmaster software for data acquisition and analyses. We have Molecular Devices Digidata 1440 and 1320 data acquisition interfaces and pClamp8.2 and 10.1 versions of the software which recognize most commercially manufactured patch clamp amplifiers for the purpose of telegraphing the amplifier gain and filter settings.

8. Recording chamber and bath reference: The easiest recording chamber to use is a 35 mm diameter culture dish filled with 3–4 mL of BSS. Glass bottom dishes or coverslips (#1 thickness, 0.13–0.16 mm thick, 12 or 25 mm diameter) coated with poly-l-lysine or an extracellular matrix protein, such as fibronectin, laminin, will be required for quantitative fluorescence measurements. The outer rim of the cell culture dish lid glued to the insert of the microscope stage is cheapest way to affix a culture dish to the microscope stage to prevent the dish from moving during the experiment. Since an Ag/AgCl wire is used to make electrical contact of the patch electrode with the patch clamp amplifier, an Ag/AgCl junction should be used to make electrical contact with the bath chamber filled with BSS. One cannot place an Ag/AgCl wire directly in the bath solution, since Ag^{2+} ions will leach into solution and kill the cells over a period of minutes. Thus, we use an agar bridge fashioned from a glass capillary tube, such as a hematocrit tube, filled with BSS and 1 % agar and stored in BSS at 4 °C until the day of use. One end of the bridge is placed in the cell culture dish or recording chamber and the other end is placed into an external reservoir filled with 1 mL of IPS containing an Ag/AgCl pellet as a reference electrode. We use 2 mm × 4 mm diameter Ag/$AgCl_2$ pellets (WPI) with a 5 cm wire centrally located as our reference electrode. The patch cords supplied with the amplifier head stage are connected to the reference electrode to establish the external ground reference for the patch clamp recordings. Thus, the bath solution is connected to the external (i.e., ground) reference electrode and the amplifier head stage by IPS-Ag/AgCl wire junctions to minimize the voltage offset between the bath and the patch electrodes.

9. Oscilloscope: A digital storage oscilloscope is somewhat optional given that most acquisition software packages have an

oscilloscope feature, but an independent oscilloscope is useful when first measuring the patch pipette resistance in the bath, forming the GΩ seal, and rupturing the membrane patch to achieve the whole cell configuration. Once the patch clamp recording is initiated, the experimenter will primarily be involved with computerized data acquisition of the gap junction voltage clamp protocols being applied to the cell pair.

10. Voltage stimulator: Again, this is an optional feature with the existence of today's computerized data acquisition and software packages, since the data acquisition interfaces also operate as a digital-to-analog (D/A) converter capable of converting computer generated voltage clamp protocols into analog signals to be applied to the whole cell recording via the patch clamp amplifier. We still rely on a custom-built voltage stimulator with two independent voltage and transistor-transistor logic (TTL) trigger outputs for our dual whole cell recordings of gap junctional currents and conductance measurements. Commercially manufactured multichannel voltage stimulators are available, such as Panlab LE12000 series, WPI Pulsemaster or A310 Accupulser signal waveform generator.

11. Fluorescence illumination system: In order to perform fluorescent dye transfer experiments or view fluorescent reporters for transiently transfected cells, one will need an epifluorescence illumination system. There are numerous fluorescence illumination systems available and the choice is usually determined by the microscope manufacturer or compatibility with the patch clamp electrophysiology software. Popular epifluorescence excitation (ex) and emission (em) filter sets (in nm) are fluorescein isothiocyanate (green) 480/40ex–535/50em, tetramethylrhodamine (red) 540/25ex–605/55em, enhanced green fluorescent protein (green) 470/40ex–525/50em, and, for dye transfer, Lucifer Yellow 425/40ex–540/50em.

12. Bessel filter: Unfiltered whole cell currents contain high frequency components that will obscure any data for analytical purposes. Thus, patch clamp current recordings are low-pass filtered and 4-pole or 8-pole Bessel filters are usually used for this purpose. Patch clamp amplifiers usually have a built-in filter with four or more settings. The Axopatch 200B has a 4-pole Bessel filter with settings of 1, 2, 5, 10 or 100 kHz. Whole cell recordings will typically be recorded at 1 or 2 kHz. This requires a digital sample rate of 200 μs or faster to prevent aliasing (i.e., distortion) of the original analog signal, resulting in artifacts of the original current signal [18]. We digitally sample at 10 kHz for a 1 kHz low-pass 4-pole Bessel filtered signal. For longer current recordings, we low-pass filter at 100, 200, or 500 Hz using a Warner Instruments LPF202A 4-pole Bessel filter and digital sample rates of 1, 2 or 4 kHz respectively.

To determine the response time of our whole cell recording apparatus, we measured the rise times in response to an instantaneous voltage step using a model whole cell circuit at filter settings of 100, 200, 500, 1000, 2000, and 5000 Hz and selected the pClamp sample interval that was twice as fast as the half-full amplitude rise time of the low-pass filtered signal.

2.3 Cultured Cells

1. Communication-deficient cell lines for exogenous connexin expression: The two commonly used cell lines for the patch clamp study of wild-type (WT) and disease mutation connexin-specific gap junctions are the mouse neuro2A neuroblastoma cell line (N2a cells) and HeLa cells (*see* **Note 4**) [12, 13]. HeLa cells tend to exhibit more endogenous coupling from Cx45 expression than N2a cells and also form larger gap junctional plaques suitable for immunocytochemical localization of WT and mutant connexins, which translates into higher g_j values that may limit the accuracy of the g_j measurements. Spherical cell geometry improves the speed and efficiency of the whole cell patch clamp and N2a cells typically exhibit a spherical cell morphology, less any neurite outgrowths, whereas HeLa cells possess a flatter, squamous cell appearance. Both cell types are easily amenable to whole cell patch clamp procedures. Both of these cell lines are typically grown in culture media consisting of minimum essential medium, 10% fetal bovine serum (FBS) (*see* **Note 5**), 1× nonessential amino acids, 2 mM l-glutamine, 100 U/mL penicillin, and 100 μg/mL streptomycin, sterile-filtered upon preparation, stored at 4 °C and kept sterile using aseptic cell culturing techniques. This cell culturing method works well for the parental and transiently transfected N2a and HeLa cells. Each T25 (i.e., 12.5 cm²) flask is typically passaged once per week and patch clamp dishes, such as 35 mm diameter culture dishes, prepared by adding $1–2 \times 10^5$ cells to each culture dish for use within the next 48 h. Culture media (i.e., 10 mL) is replaced as needed on a daily basis, typically once per week, more frequently for faster growing cells.

2. Primary cell cultures: To study the function of endogenously expressed connexins requires preparation of native (i.e., primary) cell cultures from live animal tissue, such as heart, vascular, neuroendocrine, exocrine, neuronal, and liver tissue. The dissociation of live tissues into its viable cellular components varies depending on the tissue type and amount of connective tissue requiring enzymatic digestion, such as collagenase, but the requirements for whole cell patch clamp purposes are essentially the same, namely a clean cell membrane preparation free of connective tissue and glycocalyx that will interfere in GΩ seal formation [19, 20]. Primary cells from embryonic or

neonatal tissues are generally easier to dissociate and culture in vitro since the tissues typically have less connective tissue to digest and the isolated cells are more calcium-tolerant. For neonatal mammalian cardiomyocyte cultures, we use M199 culture media supplemented with 10% FBS and 100 U and µg/mL penicillin and streptomycin [21].

3 Methods

1. Select an isolated cell pair: Under 100× or 200× magnification, move the microscope stage containing the 35 mm cell culture dish affixed to the microscope stage until a suitable cell pair is identified and centered in the field of view.

2. Fill patch electrodes with IPS: Patch electrode glass does not contain a capillary monofilament that helps to fill the tapered tip of the patch electrode, so one will want to fill the tip of patch electrode with a small volume of IPS while avoiding trapping air bubbles and then backfill the barrel (i.e., shaft) of the patch pipette to approximately halfway, being sure to not backfill the patch electrode beyond the chloride-coated portion of the Ag/AgCl wire of the patch pipette holder. If air bubbles get trapped in the tip of the electrode during filling, firmly hold the patch pipette tip down between the thumb and finger of one hand and flick the barrel of the pipette with a finger of the free hand to dislodge the bubbles until they rise into the barrel of the pipette and disappear. This is best accomplished before backfilling the barrel of the pipette with IPS to the midpoint. We use nonmetallic 28G (0.25/0.35 mm ID/OD) MicroFil needles from WPI for filling our patch electrodes since metal ions may leach from metallic syringe needles into the IPS during use. Insert the backfilled patch electrode onto the pipette holder, tighten and swing into position over the culture dish. Apply positive pressure, approximately 4 in. or 10 cm, using a manometer filled with colored water to measure the height of the air pressure.

3. Measure the patch electrode resistance (R_{el}): Carefully center each patch electrode in the field of view just above the bath surface and to each side of center to prevent the tips of electrodes from colliding and breaking off the electrode tips. Lower one electrode at a time into the bath while applying a low voltage pulse to each electrode until electrical contact is achieved. We use a 10 ms, +200 µV voltage step from 0 mV to measure R_{el}. The ideal value of R_{el} is 4–5 MΩ, which corresponds to 40–50 pA of current. Larger patch electrodes (i.e., 2–3 MΩ and 60–100 pA) are more difficult to form stable GΩ seals and smaller patch electrodes (i.e., >6 MΩ and <30 pA)

tend to have higher access resistances after patch break and may be more difficult to rupture the membrane patch to achieve the whole cell recording configuration. Null any patch electrode offset using the pipette offset (i.e., potentiometer) knob on the front panel of the amplifier *prior to* GΩ seal formation.

4. Form the GΩ seal: Position the patch electrodes above and to the right and left of the selected cell pair using the coarse and then fine control XYZ micromanipulators and switch to 400× magnification. While focused on the top of the cell pair, move each patch electrode over the respective cell for each patch electrode to record from, preferably just to the right and left of center, respectively. Thus, the patch electrode should first come into close contact with the cell by vertically lowering the patch electrode towards the upper surface of the cell, preferably at a tangent to the spherical surface of the cell. The opening at the tip of the patch electrode should actually not touch the surface of the cell or anything else until the positive pressure is released and negative pressure is applied in a single continuous motion to suck the cell membrane onto the end of the patch electrode, forming the GΩ seal. With any luck, one will be able to visualize the slight cupping of the cell membrane resulting from the positive pressure on the patch electrode and resulting IPS stream *prior to* reversing the pressure on the patch electrode to form the GΩ seal. Use as little negative pressure as necessary to form the GΩ seal, typically 2–5 cm H_2O, with a clean electrode and cell membrane. Excess pressure may draw too much of the cell membrane into the tip of the patch pipette, prematurely rupture the membrane patch covering the opening of the electrode or blow the GΩ seal. Gently release the negative pressure once the GΩ seal is formed. Repeat these steps with the second electrode. While applying the 200 μV pulse, the current trace will flatten out to a straight line. To measure the GΩ seal, a 5, 10 or 20 mV will have to be applied to observe the seal current. A value of 1 pA/mV equals 1 GΩ of resistance.

5. Compensate the patch electrode capacitance: Compensate the capacitive current transient during the mV voltage pulse used to measure the seal resistance by setting the Bessel filter to >10 kHz or wide band (i.e., bypass) and using the fast and slow capacitance compensation circuits on the front panel of the patch clamp amplifiers to minimize the capacitive transient arising during the onset of the voltage pulse. The capacitive transient will vary slightly between patch electrodes owing to the thickness of the pipette tip submerged in the bath solution. Typically, only minor adjustments of the fast capacitance compensation circuit will be required between patch electrodes provided the geometry does not vary significantly.

6. Rupture the membrane patch: The whole cell recording configuration is usually achieved by applying negative pressure to the patch pipette until the cell membrane patch covering the opening of the electrode ruptures, establishing access to the cell interior and permitting voltage clamp control of the cell membrane potential (V_m). An alternative to the negative pressure (i.e., suction) ruptured patch approach to achieve the whole cell configuration is the perforated patch technique wherein a perforating agent is added to the patch electrode to permeabilize the membrane patch to ions, thus establishing electrical contact, but not small molecules like fluorescent dyes and second messengers, such as cyclic adenosine monophosphate (see **Note 6**).

7. Measure the whole cell electrode resistance (R_{el}): Determining the value of R_{el} is critical if one wants to apply corrective measures to the measurement of g_j. R_{el} will increase as a result of rupturing the cell membrane patch and is calculated by measuring the time constant of the whole cell capacitive current decay (τ_c) and cellular input capacitance (C_{in}) using the equation $R_{el} = \tau_c / C_{in}$ [22]. Calculation of the electrode series resistance is an automatic feature of some data acquisition programs, like pClamp, if one is using their built-in D/A converter to generate voltage clamp protocols. We routinely use a 5 mV step from a holding potential (V_h) of −40 mV to −35 mV for 10 ms and signal average 10 capacitive current (I_c) transient signals to fit the transient with an exponential decaying function $I_c = I_0 \cdot e^{-t/\tau_c} + C$, where I_0 is the peak amplitude of the capacitive current transient, τ_c is the decay constant and C is the steady-state whole cell current value at the new voltage (−35 mV). Since $C_{in} = \Delta Q / \Delta V$, integrating the area under the capacitive transient curve ($I = \Delta Q / \Delta t$, so $I \times \Delta t = \Delta Q$) provides a measure of the amount of charge required to charge the cell capacitance to the new V_m (ΔQ in pA/ms) and dividing that value by 5 mV yields the value of C_{in} (Fig. 1). Once τ_c and C_{in} have been calculated for each cell, then R_{el} is calculated for each electrode. The 5 mV step in voltage must be simultaneous applied to both cells, otherwise a transjunctional voltage (V_j) gradient will result and an unwanted junctional current (I_j) component will be added to the whole cell current signal (i.e., C will change in opposite directions in both cells).

8. Assessment of the whole cell input resistance (R_{in}) and gap junction conductance (g_j): Knowing the value of R_{in} is not necessary to measure g_j, but is helpful in determining the range of V_m to use during the V_j voltage clamp protocols to be applied during the experiment. A high R_{in} value keeps the nonjunctional membrane currents (I_m) to a minimum and improves the I_j signal-to-noise ratio. R_{in} can be determined by any ΔV_m step

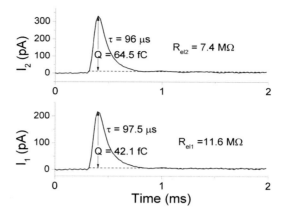

Fig. 1 Measurement of whole cell patch electrode (access) resistance (R_{el}). After rupturing the membrane patch of both GΩ-sealed electrodes, a +5 mV voltage command (V_c) step is applied simultaneously to both cells of a coupled cell pair from a common holding potential (V_h) of −40 mV and the whole cell capacitive current transients (i.e., I_1 and I_2) are recorded using the wide-band (i.e., 100 kHz or filter bypass) setting on the 4-pole Bessel filter. Fitting the decay phase with an exponential function yields the decay time constant, τ_c, which is equal to the product of the whole cell capacitance (C_{in}) × R_{el}. Thus, after integrating the area under the I_1 or I_2 curve to obtain the value of C_{in} (= Q/V = (pA·ms)/5 mV), R_{el} is calculated based on the expression $R_{el} = \tau_c/C_{in}$ [22]. These signals were digitally sampled at 50 kHz

applied simultaneously to both cells, since $R_{in} = \Delta V_m/\Delta I_m$, but this only determines the value of R_{in} at 1 V_m. Creating an I_m–V_m current–voltage relationship over a 200 mV range of potentials requires only a few seconds and is easily accomplished by running a V_m ramp from −140 mV to +60 mV with a slope of 20 ms/mV = 4000 ms total duration (Fig. 2). Applying the same ramp to one cell of the pair provides a rapid assessment of the g_j value and whether V_j gating is observable or not (Fig. 3).

9. Perform the modulation of the gap junction conductance experiment: The simplest approach to measuring g_j is to apply a voltage step (ΔV) to one cell while holding the V_m of the partner cell constant at a common V_h (i.e., 0 or −40 mV). A small amplitude pulse (i.e., 10 or 20 mV) is preferable because higher amplitude ΔV steps may induce V_j gating, causing I_j and g_j to change during the pulse. The action of a pharmacological modulator of g_j, such as chemical gating by pH, Ca^{2+}/calmodulin, phosphorylation, gap junction blockers, like carbenoxolone, or gap junction agonist, such as rotigaptide [23], is easily monitored by applying the small amplitude ΔV step periodically (i.e., every 15, 30 or 60 s) during application and washout, if any, of the g_j modulator being investigated. Fluorescent dye

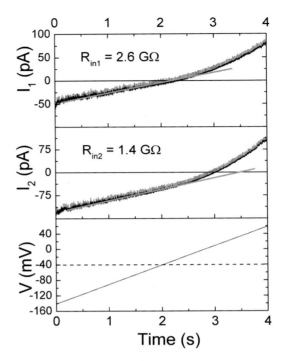

Fig. 2 Measurement of whole cell input resistance (R_{in}). Simultaneous application of a ±100 mV, 20 ms/mV voltage ramp to both cells of a coupled cell pair produces a whole cell membrane current (i.e., I_1 and I_2)–voltage relationship illustrating a linear R_{in} at negative potentials in this pair of HEK293 cells. From Fig. 1, the ratio of $R_{el}/R_{in} \times 100$ is 0.45 % for cell 1 and 0.53 % for cell 2. Thus, while operating within the linear R_{in} range for these cell pairs, the series resistance error for controlling V_m of each cell is only 0.5 % (i.e., 500 μV/100 mV). These signals were low-pass filtered at 500 Hz and digitally sampled at 4 kHz

transfer is readily assessed by adding a known concentration of a gap junction permeable fluorescent dye to one patch pipette, normally the electrode to which the ΔV step is being applied, and monitoring change in fluorescence of the recipient cell. Thus, dye permeability may be directly correlated with g_j, provided that the dye concentrations in the donor and recipient cells are known for each time step (ΔV pulse) [24]. A $V_j = 0$ mV rest interval of at least equal to the duration of the ΔV pulse (i.e., 50 % duty cycle) should be included between each test pulse. The baseline whole cell currents (I_1, I_2) should remain constant for the duration of the experiment if stable dual whole cell recording conditions (i.e., constant R_{in}) are maintained. A nonzero V_h, centered in the high R_{in} range of V_m determined by the simultaneous V_m ramp is preferable since any change in R_{in} will be detected as an increase in I_1 or I_2 during the $V_j = 0$ mV baseline interval. Since the bath potential is 0 mV by

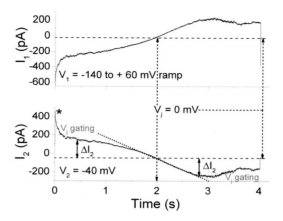

Fig. 3 Rapid assessment of junctional conductance (g_j). Application of same voltage ramp illustrated in Fig. 2, while keeping V_h constant at −40 mV in cell 2 produces a ±100 mV V_j ramp. Now the I_1 and I_2 signals will have opposite sign. I_1 this time will contain the same nonjunctional membrane current response shown in Fig. 2 plus an I_j component resulting from $V_1 \neq V_2$. Since V_2 is constant at −40 mV, the difference between the baseline I_2 current value (dashed line, when $V_1 = V_2 = -40$ mV) and the I_2 curve represents ΔI_2, the equivalent of the I_j current component from cell 1 being subtracted out of cell 2 in order to maintain V_{m2} constant. Thus, the unilateral application of the voltage ramp allows for a rapid assessment of g_j from the initial peak (*asterisk*) of I_2 ($\approx$450 pA) ÷ 100 mV $\cong$ 4.5 nS and the observation that V_j-dependent gating is evident in this HEK293 clone 80-1 (i.e., stable Cx43 short hairpin-based RNA interference knockdown [33]) cell pair. Note that the additional 450 pA of whole cell current at the start of the V_1 ramp will result in a 3 mV drop across each whole cell patch electrode, resulting in about 6 % error in V_j that was not present when ramping both cells simultaneously. High (GΩ) R_{in} cells with $g_j < 10$ nS to keep are required to maintain 90 % accuracy in uncorrected g_j measurements using the dual whole cell patch clamp technique

definition (i.e., external reference), holding at $V_h = 0$ mV will not result in a detectable change in the I_1 or I_2 baseline during the $V_j = 0$ mV rest intervals between ΔV pulses, only during the ΔV step during which g_j is measured, which may lead to errors in the g_j calculation.

10. Measurement of g_j: Since $R = V/I$, the simplest calculation of g_j is $g_j = I_j/V_j$ since $g = 1/R$. Transjunctional voltage (V_j) is defined as the membrane potential (V_m) difference between the two cells and the simplest calculation of V_j is to use the difference in the voltage clamp steps (i.e., V_1 and V_2) applied to both cells. Initially, V_1 should equal V_2 ($= V_h$) and a ΔV step is then applied to one cell of the pair, which we will define as cell 1 (i.e., the prejunctional cell). Thus $V_1 = V_h + \Delta V_1$ and $V_2 = V_h$. V_j may be defined as $V_1 - V_2 = \Delta V_1$ or $V_2 - V_1 = -\Delta V_1$. The ΔV_1 pulse

is the source of I_j, which will flow out of cell 1 into cell 2 (i.e., the postjunctional cell), resulting a change in I_2 (ΔI_2) during the ΔV_1 step. The ΔV_1 step and ΔI_2 response will always have opposite sign, as will I_1 and I_2 (i.e., the equal magnitude and opposite polarity criterion [7]), since, by design, the cell 2 patch clamp amplifier will subtract out the I_j current flowing into cell 2 in order to keep V_2 constant. Thus, $\Delta I_2/\Delta V_1 = I_j/V_j$ will result in a negative g_j value unless the sign of one of the values is reversed. Hence, if ΔI_2 is taken as the value of I_j, then V_j has to equal $V_2 - V_1 = -\Delta V_1$. Conversely, if ΔV_1 is taken as the value of V_j, then I_j has to equal $-\Delta I_2$. Realistically, V_m does not equal the command voltage (V_c) applied to each cell, since the whole cell current flowing across the whole cell patch electrode resistance for each cell will result in a voltage drop across the electrode or $V_m = V_c - I_m \cdot R_{el}$ for each cell [22]. Thus, $V_{m1} = V_1 - I_1 \cdot R_{el1}$ and $V_{m2} = V_2 - I_2 \cdot R_{el2}$. If I_j is chosen to equal $-\Delta I_2$ since I_j is being subtracted from the postjunctional cell to maintain V_2 constant, then the accurate g_j calculation is

$$g_j = \frac{-\Delta I_2}{\left[\left(V_1 - I_1 \cdot R_{el1}\right) - \left(V_2 - I_2 \cdot R_{el2}\right)\right]}$$

$$= \frac{-\Delta I_2}{V_1 - I_1 \cdot R_{el1} - V_2 + I_2 \cdot R_{el2}} = \frac{-\Delta I_2}{V_{m1} - V_{m2}} \cong \frac{-\Delta I_2}{\Delta V_{m1}} \approx \frac{-\Delta I_2}{\Delta V_1}$$

after correcting for series resistance errors [25]. This correction method makes significant differences in the calculation of the half-inactivation voltage (V_0 or $V_{1/2}$) of g_j in response to increasing V_j, called fast V_j gating, defined by the steady state $g_j - V_j$ relationship as described by the Boltzmann distribution

$$g_{j,\infty} = \left[\frac{g_{j,max} - g_{j,min}}{\left(1 + \exp\left[A\left(V_j - V_0\right)\right]\right)}\right] + g_{j,min} \text{ where } A = (nq/kT) = (zF/RT)$$

or when calculating the electrical distance (δ) to the V_j-dependent site of ionic block within a channel pore

$$K_m\left(V_j\right) = \left(\frac{b_{-1}}{b_1}\right) \cdot \exp\left(\frac{-zF\delta V_j}{RT}\right) \text{ originally derived by Woodhull}$$

[3, 26, 27]. Steady-state g_j may be normalized by dividing $g_{j,\infty}$ by $g_{j,max}$, yielding the expression

$$G_j = \frac{g_{j,\infty}}{g_{j,max}} = \left[\frac{1 - g_{j,min}}{\left(1 + \exp\left[A\left(V_j - V_0\right)\right]\right)}\right] + \frac{g_{j,min}}{g_{j,max}} \text{, where}$$

$$G_j = \frac{g_{j,min}}{g_{j,max}} \text{ will vary between 0 and 1. To minimize the series}$$

resistance errors associated with the dual whole cell recording, measurements of g_j, I_1 and I_2 need to be minimized (i.e., the lower the g_j, the higher the accuracy).

These procedures will lead to accurate measurements of V_j-dependent processes of connexins and alleviate any confusion about the assignment of the polarity of V_j gating based on the conventions used to define g_j.

11. Measuring single gap junction channel conductance (γ_j): The conductance of a single gap junction channel (γ_j) is determined by measuring the current amplitude (i_j) of a single gap junction channel opening and dividing this value by V_j. Resolving single gap junction channel openings typically occurs in poorly coupled cells with $g_j < 1$ nS and fewer than six channels, so the correction methods defined above are not essential. The I_1 and I_2 currents will be small, tens of pA and so will the voltage drop across each electrode (i.e., 10 pA×10 MΩ = 100 µV and 100 µV/40 mV×100 = 0.25 % error). It is usually customary to determine the i–V relationship for an ion channel and developing an i_j–V_j relationship provides the most accurate measure of γ_j, especially when working with rectifying gap junction channels, such as connexin hemichannels or heterotypic gap junction channels, or using asymmetric ionic solutions to measure the relative ionic permeability of connexin-based channels [28].

4   Notes

1. Approximately 2 cm of bare silver wire should protrude from the open end of the patch electrode holder. This segment of the silver electrode wire will need to chloride electroplated (chlorided) *prior to* use. Clean the bare wire (i.e., light sanding with fine grit sand or emery paper to remove any tarnish) and immerse the distal 2 cm of wire in 3 M KCl solution. One will need a second wire or a Ag/AgCl$_2$ pellet to serve as the cathode (i.e., negative voltage terminal) for the electroplating process. Attach the electrode wire to be chlorided to the positive (i.e., red) terminal of a low amperage voltage source and the Ag/AgCl$_2$ pellet to the negative (i.e., black) terminal of the voltage source. A DC power supply with variable current (i.e., mA) and voltage (i.e., 0–12 V) voltage outputs or a 9 V transistor battery will suffice as the voltage source. The current flow should be <1 mA, such as 0.15 mA for 2 cm of 0.25 mm diameter wire [29]). The chlorided wire should turn light gray in color during the electroplating process after a few minutes. Rinse with distilled water and reassemble the patch pipette electrode holder for use. An alternative approach is to immerse the clean Ag wire in Clorox bleach for a few minutes.

2. Hydraulic micromanipulators are typically more prone to drift, especially under heavy mechanical load than piezo-electric, stepper-motor, or direct-drive models. Many of the piezo-electric, stepper-motor, or direct-drive models now have fine submicron control of movement that closely approximates the smooth continuous motion of hydraulic micromanipulators.

3. The bright field illumination required for phase contrast or DIC imaging of live cells is powered by 50 or 60 Hz AC, 110 V power source and will transmit electrical noise onto the microscope stage, the patch electrodes and the recording chamber. Typically, one will have to use a low-noise DC power source with variable 12 V output to provide bright field illumination for contrast imaging of the cells to be patch clamped.

4. HEK293 cells have been reported to be communication-deficient although there are published reports to the contrary [30–33]. After performing dual whole cell patch clamp experiments on HEK293 cells from three different sources over the years and attaining high saturating levels of g_j each time, we conclude that HEK293 cells are not communication-deficient and low levels of Cx43-based coupling are evident even after successful stable short hairpin-based RNA interference knockdown of Cx43 expression and g_j by 75–90 % [33].

5. There are multiple vendors and choices of FBS for cell culturing purposes and many of them will permit stable cell growth for culturing purposes. However, we have found that some FBS lots that are perfectly acceptable for cell culturing needs are less than optimal for patch clamp purposes (i.e., poor GΩ seal formation, cells too fragile or rigid). Thus, we purchase small samples of FBS lots and screen the different FBS lots by patch clamping N2a cells to determine the ease of GΩ seal formation and membrane rupture to achieve the whole cell patch electrode recording configuration. Once a suitable FBS lot is identified, we purchase a 3–5 year supply of that FBS and utilize that lot until the entire supply is consumed or expires after multiple years of storage at –80 °C.

6. There are several advantages of using perforated patch techniques. Achieving the whole cell patch electrode recording configuration by the conventional ruptured patch approach allows for dialysis of the intracellular milieu, which is advantageous for the introduction of a fluorescent dye or controlling the intracellular ionic concentrations or pH, but may also result in the washout of essential second messengers necessary for a physiological response to neurotransmitters or hormones, such as acetylcholine and adrenaline. Intracellular dialysis by a conventional whole cell patch electrode will also dilute the transfer of any fluorescent dye from the donor to recipient cell,

which may dramatically influence the dye permeability measurements, especially when gap junction permeability is low. Thus, using a perforated patch electrode on the recipient cell of a dye transfer/g_j experiment improves the accuracy of the molecular permeability calculation as first performed by Valiunas and colleagues [24]. Nystatin and amphotericin B are polyene antibiotics that form small pores in cell membranes small enough to allow ionic flow across the membrane patch, thus permitting electrical whole recordings, but too small for larger molecules such as fluorescent dyes and cyclic nucleotides to pass through the membrane patch [34, 35]. They are not water-soluble and a concentrated stock solution (i.e., 60 mg/mL) is prepared in dimethylsulfoxide the day of use and diluted 250 times with IPS. One may attempt to keep the concentrated stock solution stored at –20 °C for the week and dilute daily as needed. The tapered tip of the patch pipette has to be filled with normal IPS *prior to* backfilling the shaft of the patch electrode with the nystatin or amphotericin B IPS solution since the perforated patch solutions will interfere with GΩ seal formation. If the diffusion distance is too long, the permeabilization of the membrane by the perforated patch solution will take additional time beyond the ≤5 min usually required to form a GΩ seal. β-escin is a water-soluble compound and is stored at room temperature *prior to* dissolving in deionized water, making β-escin easier to work with than nystatin or amphotericin B [36]. When dissolved at 30–50 μM in IPS, the patch electrode may be filled the conventional way without concern about interfering with GΩ seal formation [24]. A 50 mM stock of β-escin in deionized water may be stored at –20 °C for 1 week and diluted 1000–2000 times with IPS for daily use to yield a working concentration of 50 μM or less.

Acknowledgements

This work was supported by NIH grant HL-042220 and a Hendricks Fund grant to R.D.V. Xian Zhang and Dakshesh Patel proofread the manuscript.

References

1. Auerbach AA, Bennett MVL (1969) A rectifying electrical synapse in the central nervous system of a vertebrate. J Gen Physiol 53: 211–237

2. Spray DC, Harris AL, Bennett MVL (1979) Voltage dependence of junctional conductance in early amphibian embryos. Science 204: 432–434

3. Spray DC, Harris AL, Bennett MVL (1981) Equilibrium properties of a voltage-dependent junctional conductance. J Gen Physiol 77: 77–93

4. Harris AL, Spray DC, Bennett MVL (1981) Kinetic properties of a voltage-dependent junctional conductance. J Gen Physiol 77: 95–117

5. Hamill OP, Marty A, Neher E et al (1981) Improved patch-clamp techniques for high-resolution current recording from cells and cell-free membrane patches. Pflugers Arch 391:85–100

6. Neyton J, Trautmann A (1985) Single channel currents of an intercellular junction. Nature 317:331–335

7. Veenstra RD, DeHaan RL (1986) Measurement of single channel currents from cardiac gap junctions. Science 233:972–974

8. Paul DL (1985) Molecular cloning of cDNA for rat liver gap junction protein. J Cell Biol 103:123–134

9. Beyer EC, Paul DL, Goodenough DA (1987) Connexin43: a protein from rat heart homologous to a gap junction protein from liver. J Cell Biol 105:2621–2629

10. Dahl G, Miller T, Paul D et al (1987) Expression of functional cell-cell channels from cloned rat liver gap junction complementary DNA. Science 236:1290–1293

11. Barrio LC, Suchyna T, Bargiello T et al (1991) Gap junctions formed by connexins 26 and 32 alone and in combination are differently affected by applied voltage. Proc Natl Acad Sci U S A 88:8410–8414

12. Hennemann H, Suchyna T, Lichtenberg-Fraté H et al (1992) Molecular cloning and functional expression of mouse connexin40, a second gap junction gene preferentially expressed in lung. J Cell Biol 117:1299–1310

13. Veenstra RD, Wang HZ, Westphale EM et al (1992) Multiple connexins confer distinct regulatory and conductance properties of gap junctions in developing heart. Circ Res 71:1277–1283

14. Bers DM, Patton CW, Nuccitelli R (2010) A practical guide to the preparation of Ca^{2+} buffers. Methods Cell Biol 99:1–26

15. Flaming DG, Brown KT (1982) Micropipette puller design: form of the heating filament and effects of filament width on tip length and diameter. J Neurosci Methods 6:91–102

16. Rae JL, Levis RA (2004) Fabrication of patch pipets. Curr Protoc Neurosci 26:1–32

17. Sutter Instrument Company (2015) Pipette cookbook 2015. P-97 and P-1000 micropipette pullers. Rev E. http://www.sutter.com/PDFs/pipette_cookbook.pdf

18. Colquhoun D, Sigworth FJ (1983) Fitting and statistical analysis of single-channel records. In: Sakmann B, Neher E (eds) Single-channel recording. Plenum Press, New York, pp 191–263

19. Trube G (1983) Enzymatic dispersion of heart and other tissues. In: Sakmann B, Neher E (eds) Single-channel recording. Plenum Press, New York, pp 69–76

20. Spector I (1983) A primer in cell culture for pathologists. In: Sakmann B, Neher E (eds) Single-channel recording. Plenum Press, New York, pp 77–90

21. Lin X, Gemel J, Beyer EC et al (2005) A dynamic model for ventricular junctional conductance during the cardiac action potential. Am J Physiol Heart Circ Physiol 288:H1113–H1123

22. Marty A, Neher E (1983) Whole-cell recording. In: Sakmann B, Neher E (eds) Single-channel recording. Plenum Press, New York, pp 107–122

23. Lin X, Zemlin C, Hennan J et al (2008) Enhancement of ventricular gap junction coupling by rotigaptide. Cardiovasc Res 79:416–426

24. Valiunas V, Beyer EC, Brink PR (2002) Cardiac gap junction channels show quantitative differences in selectivity. Circ Res 91:104–111

25. Veenstra RD (2001) Voltage clamp limitations of dual whole cell recordings of gap junction current and voltage recordings. I. Conductance measurements. Biophys J 80:2231–2247

26. Woodhull A (1973) Ionic blockage of sodium channels in nerve. J Gen Physiol 61:687–708

27. Musa H, Veenstra RD (2003) Voltage-dependent blockade of connexin40 gap junctions by spermine. Biophys J 84:205–219

28. Veenstra RD (2001) Determining ionic permeabilities of gap junction channels. In: Bruzzone R, Giaume C (eds) Methods in molecular biology, vol 154, Connexin methods and protocols. Humana Press, New Jersey, pp 293–311

29. Warner Instruments. (2004) Chloriding Ag/AgCl electrodes. http://www.warneronline.com/Documents/uploader/ChloridingAgAgClelectrodes(2004.02.02).pdf

30. Stong BC, Chang Q, Ahmad S et al (2006) A novel mechanism for connexin26 mutation linked deafness: cell death caused by leaky gap junction hemichannels. Laryngoscope 116:2205–2210

31. McSpadden LC, Kirkton RD, Bursac N (2009) Electrotonic loading of anisotropic cardiac monolayers by unexcitable cells depends on connexin type and expression level. Am J Physiol Cell Physiol 297:C339–C351

32. Sroka J, Czyz J, Wojewoda M et al (2008) The inhibitory effect of diphenyltin on gap junctional intercellular communication in HEK293 cells is reduced by thioredoxin reductase 1. Toxicol Lett 183:45–51

33. Patel D, Zhang X, Veenstra RD (2014) Connexin hemichannel and pannexin channel electrophysiology: how do they differ? FEBS Lett 588:1372–1378

34. Horn R, Marty A (1988) Muscarinic activation of ionic currents measured by a new whole-cell recording method. J Gen Physiol 92:145–159

35. Rae J, Cooper K, Gates P et al (1991) Low access resistance perforated patch recordings using amphotericin B. J Neurosci Methods 37:15–26

36. Fan JS, Palade P (1998) Perforated patch recording with beta-escin. Pflugers Arch 436:1021–1023

INDEX

A

Adherent cell culture 157, 158, 177
Agar...217
Antibody.......................................39, 41, 44, 47–50, 55–58,
   60–67, 102, 107, 110, 115, 120, 123, 125–131
Artery 74, 76, 77,
   84, 102
Automation 146, 150, 151

B

Biotin switch 100, 102, 103,
   106–107
Bisulfite-specific PCR (BSP) sequencing...............23, 28–32
Bovine corneal endothelial cells (BCEC) 204–207,
   209, 210

C

Caged
   coumarin.....................................182, 187, 189
   dye 158, 182
Calcein 157, 173–176, 178, 184,
   185, 187, 189, 190
Calcium
   imaging..183, 203–210
   wave propagation.....................................203–210
Cardiomyocyte 146, 220
Cell culture 6, 39, 41, 46, 47, 59, 73,
   74, 78, 81, 84, 86–90, 93, 94, 102, 104, 109, 114, 116,
   126, 127, 135, 136, 140, 141, 147, 148, 158, 162, 166,
   173, 174, 178, 184, 195–197, 207–209, 217, 219, 220
Chronic atrophic gastritis 22, 29
Conductance
   gap junction.......................................213–229
   single channel 214
Confocal microscopy 136, 138, 148,
   174, 207, 209
Connexin (Cx)
   connexin 26 ....2, 4, 10, 16, 22, 38,
   41, 47, 50, 84, 86
   connexin 31 .... 84–86, 90
   connexin 31.1 .... 84–86, 90
   connexin 32 ....1, 2, 4, 10, 16, 21–23,
   26–29, 34, 38, 47, 49, 50, 58, 67, 98, 114–116, 121,
   126, 130
   connexin 43 ....2, 4, 10, 16, 21–23,
   26–29, 32, 34, 38, 47, 49, 50, 58, 64, 84, 86, 91,
   97–110, 113, 155, 163–165, 183, 186, 189, 193–201,
   205, 206, 225, 228
Connexon 56, 57, 145, 204

D

DNA methylation 1, 21–35
Dye
   coupling 133
   transfer................................133–143, 164, 182, 183, 186,
   188, 190, 218, 224, 229
Dysplasia 22, 29

E

Electroporation...74, 77, 80, 146,
   155–167, 204
Embryo.. 83, 84, 182
Embryonic development 182
Endogenous immunoprecipitation 64, 114, 219
Endothelial cell..............................72–74, 76–78, 80, 97, 98,
   100, 102, 108, 165, 206, 208–210
Endothelium 98, 100
Enhanced chemiluminescence............................... 38, 41, 44
Epigenetic 1, 22
Extraembryonic tissue 84
Ex vivo.. 73, 76–78, 135

F

Fluorescence based Western blotting........................ 100, 104
Fluorescence imaging 178, 181–190
Fluorescence recovery after photobleaching (FRAP) ....
   171–179, 204
Fluorescence recovery curve 175
Fluorophore...56, 57, 67, 183, 194
Formalin-fixed paraffin-embedded tissue
Frozen tissue... 8, 55, 58–61, 66

G

Gap junction (GJs) 0, 21, 38, 56, 84, 86,
   91, 97, 98, 100, 113–131, 134, 135, 138, 140, 141,
   145–153, 155–167, 171–179, 181–190, 193, 194,
   198, 200, 203–210, 213–229
Gastric carcinoma..22

Mathieu Vinken and Scott R. Johnstone (eds.), *Gap Junction Protocols*, Methods in Molecular Biology, vol. 1437,
DOI 10.1007/978-1-4939-3664-9, © Springer Science+Business Media New York 2016

Gene expression.. 4, 13, 14, 72,
    84–85, 200–201
Gestation 84, 86

H

Hemichannel..38, 56, 98, 156,
    203–210, 227
High frequency oscillations 158
High-throughput.. 135, 146, 152

I

Immunoblot analysis 37–52, 91, 129
Immunocytochemistry...63–64
Immunohistochemistry 55–68
Incision loading 135
Integral membrane protein......................................114, 129
Interactome 113–131
Intercellular communication.........................21, 56, 151, 156,
    181, 193, 194, 203, 204, 206
Intestinal metaplasia...22, 29
In vitro..55, 59, 84, 86–88,
    143, 157, 220

L

Labor..83–94
Liver 1–17, 57, 58, 67, 114, 117, 119,
    125, 126, 128, 203, 219
Local activation of a molecular
    fluorescent probe..204
Luciferase reporter assay................................. 197, 198, 200
Lucifer yellow 134, 136, 151,
    157, 218

M

MassArray 23, 26, 32–34
Mass spectrometry....................................32, 37, 38, 113–131
Mechanical stimulation 203–210
Membrane
    fractionation 115, 117–121
    microdomain...113–131
    preparation.. 100, 105–106, 219
    stripping 44, 49
Messenger RNA..2, 5, 10, 11, 14,
    72, 78–80
Methylation-specific PCR....................................23, 26–29
Microinjection..................................135, 145–153, 157,
    158, 172, 204
Minimum information for publication of quantitative
    real-time PCR experiments 2, 13, 17
Mobile fraction percentage...176
Monolayer 88, 133, 134, 138, 139, 141,
    148, 149, 157, 158, 160, 167, 176, 178, 194, 210

Morphometric analysis 138–139
Myometrium 86

N

Network analysis 114, 117, 124, 126
Non-atrophic gastritis 22, 29
Normal gastric mucosa 22, 29

O

Opti-Prep™ gradient...114
Osteoblast... 193, 196, 198, 200

P

Pannexin..206
Parachute assay 193, 197
Patch clamp
    perforated 222, 228, 229
    whole cell................................ 172, 204, 213–229
Phosphorylation 49, 97–110, 223
Photoactivation 182
Placenta 83, 84, 86
Pluronic 184, 189
Posttranslational modification 156
Pregnancy 83–94
Primary cells
    hepatocytes 50, 65
    myocytes 92
    vascular cells 97–110
Promoter-reporter assay 195–198, 200–201
Protein
    extraction...39, 41–42, 45, 46, 49
    quantification.. 39, 42, 45
    transfer.. 41, 43, 48
Proteomics...130

R

Reporter
    assay... 195–198, 200
    dye 157, 158, 165, 166
Resistance
    electrode 220, 222, 226
    input 213, 214, 222, 224
    membrane...221
Reverse transcription quantitative real-time
    polymerase chain reaction
    complementary DNA 2
    data analysis...2
    real-time polymerase chain reaction 1–17
    RNA extraction 2
Rhodamine-dextran.. 134–136, 148
RNA interference 71–73, 78, 225, 228

S

Scalpel loading..136
Scrape loading 133–143, 158, 172, 204
Second messenger.....................................193–201, 222, 228
Short hairpin RNAs...72
Small interfering RNA................................ 71–81, 194, 198,
    200, 205, 206
Smooth muscle cell.........................71–81, 89–93, 97, 98, 102
S-nitrosylation 97–110
Sodiumdodecylsulfate polyacrylamide
    gel electrophoresis...37–52
Stretch 86–92, 94

T

Tenocytes... 173, 174, 176, 178
Time-lapse imaging..........................157, 160–162, 165, 182
Tracers... 157, 158, 173, 182
Transfection...72–81, 108, 183, 184,
    195–198, 200

Trophectoderm..84
Trophoblast
    differentiation..84, 88
    stem cell...83–94
Two photon
    imaging..182
    uncaging 184, 187–188
Tyramide signal amplification 57, 58, 60,
    62, 64, 68

U

Uterine myocytes....................................... 84, 86–89, 91–93
Uterus.. 83, 86, 89

V

Vascular cells... 74–75, 86, 97–110
Voltage
    membrane.. 156, 213, 214
    transjunctional 214, 222, 225